A-Z
ANIMAL PHYSIOLOGY

A-Z ANIMAL PHYSIOLOGY

Prof. Manoranjan Tripathi

CENTRUM PRESS
NEW DELHI-110002 (INDIA)

CENTRUM PRESS

H.O.: 4360/4, Ansari Road, Daryaganj,
New Delhi-110 002 (India)
Ph.: 23278000, 23261597

B.O.: No. 1015, Ist Main Road, BSK IIIrd Stage
IIIrd Phase, IIIrd Block,
Bangalore - 560 085 (India)
Tel.: 080-41723429
Visit us at: www.centrumpress.com

A-Z Animal Physiology

First Edition, 2009

ISBN 978-93-80106-02-1

PRINTED IN INDIA

Printed at Salasar Imaging Systems, Delhi-110035 (India)

Contents

Preface

Animal Physiology deals with the experimental analysis of normal processes in normal systems in contrast to Pathophysiology that subserves medicine as attended by injury and disease. There is a clear reciprocity of fundamental insight that emerges from comparisons of normal and abnormal processes. Students of Biology do not possess an adequate background of Physics and Chemistry; as a result they face much difficulty in understanding the true concepts of Physiology.

The present textbook has been written mainly for such students, but it is sure to serve as a real guide to those as well who intend to specialize in Animal Physiology. The phenomena of life processes cannot be explained without going into the details of biochemical events that occur at molecular level; the first three chapters of this book are devoted to the structure, chemical properties and biological significance of macromolecules of physiological importance to prepare the students to understand the part of the material which follows in the subsequent chapters.

A-Z Animal Physiology has been completely updated to include the exciting advances and to present the most current information's available. The goals and distinguishing features of Animal Physiology however, remain unchanged. This book continues to be a basic text that provides an accessible introduction for undergraduate or medical students.

Author

Chapter 1

Introduction to Animal Physiology

The field of animal physiology extends the tools and methods of human physiology to non-human animal species. Plant physiology also borrows techniques from both fields. Its scope of subjects is at least as diverse as the tree of life itself. Due to this diversity of subjects, research in animal physiology tends to concentrate on understanding how physiological traits changed throughout the evolutionary history of animals.

Fig. Animal Physiology

Other major branches of scientific study that have grown out of physiology research include biochemistry, biophysics, paleobiology, biomechanics, and pharmacology. Although the various branches of human knowledge cannot be sharply delimited from each other, it has been found expedient for practical purposes to distinguish one field of study from another.

Thus chemistry, although close to physics, is set apart from it, and within the broad field of chemistry one can recognize such special subjects as inorganic chemistry, qualitative analysis, physical chemistry, etc.

Not infrequently, as knowledge expands, the older subjects become subdivided. Thus electrochemistry and colloid chemistry are offshoots from physical chemistry. From a broad philosophical point of view some of the distinctions between sciences and their various subdivisions are none too valid. New subjects appear and receive names in treatises or college catalogs, often because some specialist desires to pass on to students information which he knows best or in which he is most interested. In spite of the fact that distinctions between various sciences and their subdivisions may be more or less arbitrary, each branch of knowledge has to a greater or less extent a reason for existence.

Perhaps one can think of each scientific subject as representing an attempt to answer a question. Thus, in biological science, anatomy covers the question of form. It tells us what organisms or their parts look like. Physiology, on the other hand, is concerned with the problem of function. It attempts to describe the actions of various organs or tissues and, more than this, it attempts to explain this action in terms of known chemical and physical forces. In such explanation, moreover, there is an effort to relate the phenomena in terms of cause and effect. Obviously there are various branches of physiological science.

Because of its intimate relation to the science and art of medicine, human physiology has been studied since the very beginning of human learning. Our knowledge of the physiology of other forms of life has been focused on certain types rather than others. Insect physiology is now rather widely studied; plant physiology is an old and well-developed subject. Some groups of organisms have been neglected, and there is comparatively little information about the physiology of certain invertebrate phyla. Sometimes it is convenient to lump together the physiological knowledge of all sorts of organisms into a subject known as "comparative physiology."

The point of view of this subject is to emphasize the differences rather than the resemblances between various types of living material. Opinions may differ as to the proper definition of general physiology. To some it may be little more than a diluted vertebrate or human physiology. I should prefer to consider general physiology in the broadest possible sense: it is the science which deals with all forms of living material. It attempts to answer the question, "What is life?" The general physiologist believes that all life has something in common; his search is often for the least common denominator of living processes. That such a search may not be entirely futile is indicated by the recent history of biological thought.

Modern progress in biological science has resulted largely from the fortunate realization that living things, no matter how diverse they appear in externals, are fundamentally much alike. There is an underlying unity which has proved of tremendous importance. Cytologists learned to know cell division and chromosomes from a study of cells best suited for their purposes. Fertilization was studied first in the egg cells of marine invertebrates, and only later was the same process shown to occur in widely different types of organisms.

The fundamental facts of genetics were discovered first for plants, and modern study in this field for the most part contents itself with work on these forms which are easy to raise and which breed rapidly. And yet no one doubts that genetic science is capable of universal application. In physiology, on the other hand, the tendency has been to stress the study of vertebrate or mammalian tissues. Physiology is so close to medicine and its facts are so certain of practical application that there is a strong desire to study animals close to man. The field of mammalian or human physiology has had a marvelous development, and it would be absurd to belittle its striking achievements.

And yet, in spite of the fact that students of the physiology of mammals have accumulated a vast store of precise and valuable information, the ultimate mechanisms underlying the activity of mammalian cells and tissues are but little understood. The physiologist has been loath to accept the method of attack so fruitful in other branches of biology. If students of genetics or cytology had insisted on studying mammals or vertebrates and no other organisms, these sciences would scarcely have developed far. Great advances in theoretical biology are made by study of that type of living material particularly adapted to the problem at hand. Cells of mammals or vertebrates are often more difficult to study than those of simpler forms.

The general physiologist, in his search for fundamental information, uses whatever material he can find best suited for his purpose. It is the aim of the general physiologist to discover, in so far as possible, the nature and mechanism of living matter. This is the *raison d'être* of the science. Its students are more or less convinced that, if one could discover the essential principles which govern the activities and life of any one living cell or living material, the same principles could be used, with minor adjustments and modifications, to explain the life of practically every type of cell.

This may seem almost too much to ask, but the progress of biology has repeatedly shown the deep underlying similarity of what seem at first the most diverse manifestations of the living process. The extreme difficulty of biological research is thus to some extent compensated by the underlying unity of living materials. Although there may be a million different species of animals, there are not a million different biologies, but rather one fundamental science with variations. In general the cell is the unit of living matter, and the life of an organism is to all intents and purposes the sum total of the life of its constituent cells.

It seems clear therefore that vital machinery must be interpreted in terms of cells, and general physiology thus becomes cellular physiology. This point of view was strongly emphasized by Verworn, who states over and over again that, if physiology is to attempt fundamental explanation of vital activity, it must be a cell physiology.

There are, however, possible exceptions to this point of view. It may be argued that not all living matter is truly cellular. Thus only by a stretch of the imagination can some of the smaller bacteria be regarded as cells; and filtrable viruses which are beyond the limits of microscopic vision, and which may or may not be alive, are certainly not cells in any ordinary sense.

This argument seems to have very little force. Biologists generally realise that the cell doctrine is not absolutely true, and that there are forms of living matter which are not truly cellular. Nevertheless, the cell doctrine is true for all practical purposes, and it forms the basis for such sciences as histology, pathology, embryology, etc. Because ultramicroscopic forms of life may exist, neither the histologist nor the embryologist has seen fit to throw over the cell theory.

In the same way physicists and physical chemists cling to the second law of thermodynamics, although it is well known that there are special conditions under which the law does not hold. There is also another reason sometimes given for not identifying general physiology with cell physiology. In investigating the activity of a mass of living matter, it is often much simpler to study great numbers of cells rather than individual units. This is especially true of chemical study, for chemical properties of cells are usually additive. Thus if one is attempting to discover the change in the glycogen content of cells during activity, one can do this best by analyzing great numbers of cells.

The analysis is, therefore, conducted on tissues rather than on cells, and in this sense, from the standpoint of method or technic, the study is no longer truly cellular. But the primary interest of the investigation is the glycogen change in the individual cell, so that this argument seems rather a superficial one. At the present time, in many branches of physiology, there is an increasing tendency to study the individual cells in so far as this is possible. Muscle physiologists are giving more and more attention to the isolated single muscle fiber, and students of nerve are attempting as best they can to isolate single nerve fibers.

As a matter of fact, these recent studies of muscle and nerve physiologists have been inspired more by a desire to avoid the complexities which arise from the existence of many fibers not always in perfect time accord, rather than from any desire to study the single fiber from the viewpoint of the student of cell protoplasm. Nevertheless, they indicate a growing realization that the ultimate mechanism responsible for any form of vital activity lies inherent in the individual cells. Whether or not we regard general physiology as synonymous with cell physiology is of no great consequence. A science is whatever its devotees choose to make it. No one could be excluded from the field of general physiology because he did not think in terms of cells.

There is, however, a growing realization that an understanding of cells and cell protoplasm is essential for the interpretation of the various vital processes as they occur in animal and plant organisms. Much of the hope for the future of general physiology lies in the possibility of using simpler types of protoplasm in order to discover the general characteristics of living substance. The field of the general physiologist is as broad as all life itself. How broad this is, at the present time it is impossible to state, for there is no agreement as to the lower limits of life.

Numerous submicroscopic filtrable viruses exist, many of them causing diseases in animals and plants. But whether these viruses are alive or not is a question still being argued. True, they reproduce and some of the larger ones resemble bacteria. On the other hand, unlike bacteria they are incapable of growth on simple synthetic media and they can only exist within bacterial, plant or animal cells. Moreover, Stanley found that the virus which causes tobacco mosaic disease is a protein molecule, and since then all the plant viruses that have been studied have been found to be protein molecules.

However, although some animal viruses seem to consist wholly of protein, others like vaccinia virus contain fat and carbohydrate as well as protein. Viruses vary widely in size and shape. Some are so small that their molecular diameter is only about 25 mμ (0.000,025 mm.). The largest virus has a volume approximately 10,000 times that of the smallest. Such large viruses can be as large or larger than small bacteria. Some viruses are spherical, some are rod-shaped; one has been found to be tadpole-shaped, with a spherical head and a slender elongate tail.

In a study of these diverse, extremely tiny, living or partly living organisms or substances, much can doubtless be learned concerning the nature of life processes.

A considerable body of information has already been accumulated. General physiology is a theoretical science just as physical chemistry is. But whereas practical chemists and chemical engineers have long since learned to depend on physicochemical principles, medical scientists and students of applied human physiology and pharmacology have sometimes been slow to realise the relation of their subjects to general physiology.

It is interesting to note that a leading student of pharmacology, which is one of the most practical of biological sciences, has stated that "The manner in which drugs exert their action on cells has become one of the most important fundamental problems of physiology. Fortunately, at the present time, especially among the younger medical men, there is an ever-growing realization of the importance of basic general physiology. The primary object of the general physiologist is to interpret and explain vital phenomena. Obviously the explanations he seeks must be in terms of known scientific concepts.

As best he can, the general physiologist attempts to give a complete description of living substance in terms of its physical and chemical properties; and more than that, he attempts to show exactly what physical and chemical changes occur during vital activity. But mere description is not enough. He must aim to find causal relations between the physical and chemical phenomena observable in the living material and the life itself; in other words he is interested in discovering mechanism.

This is his most difficult task. To measure mechanical, thermal, electrical or chemical change in plant and animal tissues is not too hard, but to find out exactly which changes are primarily responsible for definite phases of vital activity is always exceedingly difficult and often, at our present stage of progress, impossible. Even in the simplest types of vital activity so many changes occur that it is hard to know which ones are intimately responsible for what is happening. Similar difficulties exist even for non-living machinery.

We know how an automobile works because it has been constructed by human hands, but if we were obliged to explain the action of an automobile by what we could see, we might be inclined to believe that its movement was due to the noise it made, the heat it produced or the breeze it generated. As scientists, general physiologists are concerned only with explanations in terms of what is known or what can be known. Whether or not life can be explained in terms of known or knowable chemical and physical forces is a philosophical rather than a scientific question and it should perhaps be left to the philosophers.

For them, and for thoughtful people generally, it is a question of tremendous interest. "Since a man must needs live before he can be a philosopher, no problem of philosophy is more fundamental than the nature of life." For many years, those who believe that eventually explanations of the nature of life processes can be obtained in terms of chemistry and physics have been called mechanists. On the other hand, those who deny that physical and chemical forces can ever provide a complete explanation have been called vitalists.

Between mechanists and vitalists, the arguments have often been bitter—among working biologists generally, the term vitalist is often used as a term of reproach—and hence in recent years there has been a semantic attempt to alter the terminology or to abandon it altogether.

But the terms are still properly understood in their original meaning, and it is doubtful if other words or other definitions of the old words will make the arguments any less bitter. To some modern vitalists, the chief characteristic of living material, the one most difficult to explain chemically or physically, is the extremely complex organization of living systems.

The distinguished general physiologist Lillie believes that living material has a "directiveness" quite different from the "randomness" of inanimate material. The mechanist is impressed with the fact that because of our newer knowledge of viruses there is no longer any sharp and clear-cut distinction between living and non-living, and he also likes to argue that vital phenomena once thought highly mysterious have now been explained, to some extent at least, in physical and chemical terms. To the hard-working physiologist, intent on obtaining at least a partway explanation of some physiological process in physiochemical terms, the validity of the mechanist point of view scarcely seems worth discussing.

He is not concerned with the ultimate truth sought by the philosopher. But, unfortunately, when some elderly biologist loses faith in physiochemical interpretation as an answer to all the ultimate problems of biology and leaves his laboratory to publish defeatist or vitalistic views in philosophical journals, these views are promptly read by the philosophers, who by and large have very little acquaintance with biological journals and biological thought and who commonly are convinced that the vast majority of biologists are vitalists. Nothing could be farther from the truth.In view of the fact that general physiology is a new science, it has never been organized in the same way that older sciences have.

There has been no exact standardization of the scope of the subject and still less is there any agreement as to how it can best be approached, or how subdivided. Anyone may choose his own way of presenting the subject. After a brief survey of the morphology of living cells there will be a discussion of their chemical make-up in so far as it is known, and this will be followed by a consideration of the physical properties of living systems.

This preliminary survey of the statics of living systems will precede a more lengthy treatment of the kinetics. In considering the activities of the living machine the first section will deal with the chemical aspect of the subject; in other words, the living metabolism. This will include such topics as nutrition, digestion, respiration, secretion and excretion. The physical aspects of vital activity will then be taken up, and the energy output of the living engine—its production of mechanical energy, heat, electricity and light—will be discussed.

The living machine can function only under certain conditions of the environment, and the effect of the environment will next be considered.

Living material can adapt itself to prolonged changes in environment, and we must therefore concern ourselves with the question of acclimatization. An outstanding characteristic of living things is their power to react in special and peculiar ways to sudden changes of the environment, that is to say, to stimulation.

This characteristic is known as "irritability," and there is a large body of information concerning various phases of it. After discussing irritability we shall conclude with a brief survey of the phenomenon of aging or senescence, and a short discussion of reproduction.

Chapter 2

Animal Morphology

Animal Morphology or anatomy, is the study of the shape, form, and structure of animals and their parts. Morphologists are interested in questions such as how the fang mechanisms of cobras differ from those of rattlesnakes or how the wings of bats evolved from the forelimbs of their ancestors.

Animals can be broadly distinguished into two major groups the nonchordates (or invertebrates) and the chordates.Invertebrates account for more than 95% of the animal species that have been identified. These are animals which do not have a vertebral column or a back bone. They are distributed into nine major phyla from protozoans to echinoderms. The chordates are animals which have a notochord on the dorsal side of the body. Most of the chordates can be described as vertebrates since the notochord in these animals is restricted to the embryonic stage and in the adults it becomes replaced by a vertebral column.

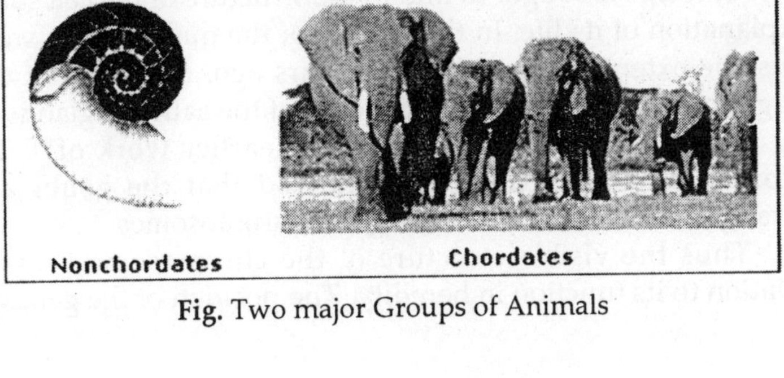

Fig. Two major Groups of Animals

A very distinct evolutionary trend in the general organisation of animals can be noticed as one proceeds from protozoans to chordates. From an acellular level of organisation to cellular level and then tissue level, organ level and organ system level of organization can be recognised. In addition, one can recognise the gradual appearance of distinct features such as symmetry, body wall and body cavity. There are further improvements in the form of appearance of segmentation, body divisions and even formation of a distinct head (cephalistion).

The culmination of all these evolutionary advances, is seen in the mammals which represent the most recent group of animals to have evolved and at the same time the most advanced group of animals. In order to gain an insight into the evolutionary advances among animals, it become necessary to study the salient features of representative animals from the group of invertebrates and vertebrates.

If one examines a living cell under the highest powers of the microscope one sees surprisingly little. In view of the fact that protoplasm is capable of complicated physical and chemical transformations it might be supposed that it would possess intricate mechanical arrangements. As everyone knows, the visible structure of a living cell is rather simple, and there is nothing to suggest the complexities of man-made machines. This may mean that the fundamental structure of protoplasm is ultramicroscopic or in the range of molecular dimensions. Following the establishment of the cell doctrine the science of cytology was built up mostly along morphological lines.

Cytologists sought to find in the structure of the cell some explanation of its life. In their study of the nucleus they were to some extent successful. Some years ago, the study of the huge chromosomes found in the cells of the salivary glands of fly larvae completely confirmed the earlier work of T. H. Morgan and his school, and showed that the genes are arranged in linear fashion along the chromosomes.

Thus the visible structure of the chromosome bears a relation to its function in heredity. The position of the genes is important in determining the way in which they are distributed under various conditions, but the fact that they are arranged in linear order tells us nothing as to how a gene can influence the development of one character or another. In their study of the structure of cytoplasm cytologists have not been so fortunate.

Although the cell nucleus is of primary importance in inheritance, for the ordinary routine life of the cell the cytoplasm plays the leading role. This is indicated by various experiments in which cells were deprived of nuclei. Thus, Lynch, among others, divided amebae into halves, one with nucleus, the other without. The fragment without a nucleus was able to live for days, although apparently it was incapable of synthesizing nitrogenous compounds. Similar studies of other workers indicate that cells without nuclei can live for a time.

A surprising case is that of Acetabularia mediterranea, a large unicellular alga which possesses but a single nucleus that is easy to remove. In this form, cytoplasm without nucleus can live for months and is capable of regenerating some rather complicated structures. On the other hand, in the nerve cells of higher animals, growth of cytoplasm can occur only in the region of the nucleus. Studies of the structure of cytoplasm have yielded very little which has proved of much help in the interpretation of its function.

Perhaps because of this, some modern cytologists seem to be losing interest in the classical cytology which dealt almost entirely with structure and have become more and more concerned with the chemistry and the physiology of the cell. All living cells are believed to be surrounded by definite membranes. The evidence for this statement has come almost entirely from the physiological side. Since the early days of cytology morphologists have been but little concerned with the outer boundaries of cells. If one examines a sea urchin egg carefully one can see that it is bounded by a delicate membrane perhaps 1μ in thickness.

This membrane is very hard to distinguish, for it appears to have the same refractive index as the sea water surrounding the egg. Whether the membrane which is visible around the sea urchin egg is the same membrane as that which governs the osmotic behaviour of the cell is not certain. The simplest theory would be to assume that it is. However, some authors insist that osmotic membranes surrounding cells are much thinner than the membrane which can be seen to envelop the sea urchin egg.

It is possible that the cell boundaries or cell walls which can be so clearly seen in preparations of animal tissues are really membranes of physiological importance. The question as to whether they are or not has scarcely been discussed. Most biologists are influenced by the relation one finds in plant tissues.

The cell walls of plant cells are non-living, and serve merely as supports for the protoplasmic structures within. On being placed in concentrated solutions plant cells shrink away from their walls. There is evidence that, within the outer membrane of the animal cell, there is a cortex, different both in morphological and physical properties from the cell interior.

In some protozoan cells such a cortex can be clearly recognized. In metazoan cells it is not so easy to distinguish. However, Just has presented both morphological and physiological evidence to show the presence of a cortex as a very important part of the metazoan cell. Within the outer membrane and the cortex of a living cell lies the main bulk of the cytoplasm. Cytologists have struggled hard to discover structures in the cytoplasm which would help to explain its activity. Because of the fact that living cells reveal relatively little, various methods of fixing, sectioning, and staining have been employed.

Cells, or groups of cells, are treated with strong acids, salts of heavy metals, alcohol or formaldehyde, or with various combinations of these and other reagents. Following fixation the cells are bathed in alcohol, then in some solvent of paraffin, and then (typically) they are impregnated with paraffin, cut into sections, and stained. These methods of treatment have been very successful in the study of chromosomes, and it can be shown by comparison with the living that the chromosomes are not markedly altered by all these procedures. The value of fixation and staining in the study of cytoplasm is less certain.

Fischer showed at great length that fixation agents caused the production of various artifacts in protein solutions, and that these artifacts resembled the granules and filaments described by cytologists as occurring in living cells. Indeed, if one watches the action of a fixing agent on living cells, one may, under proper conditions, observe the formation of new granules or vacuoles, or even the disappearance of granules normally present.

Since the publication of Fischer's book cytologists have been more careful to compare the structures they see in fixed and stained material with the living protoplasm. In the latter half of the nineteenth century students of the cell urged one type of structure or another as essential for the living protoplasm.

Thus, Frommann, Heitzmann, Klein and Leydig, all argued that protoplasm consisted of a net of solid material, the meshes of which were filled with fluid. Flemming believed that fibrils were the most important constituent of the protoplasm. On the other hand, Altmann vigourously maintained the importance of granules. He believed that the granules represented minute organisms living within a more or less inanimate fluid.

Finally, the protozoologist, Bütschli, thought of protoplasm as a foam, or rather as an emulsion composed of 2 liquids, one in the form of droplets, the other as lamellae between the droplets. These nineteenth century views of the fundamental nature of living substance are not as new as is generally supposed.

Even in the eighteenth century many observers had noted the presence of innumerable granules in living material, and there was also some talk of the existence of fibrils. The twentieth century saw a new development, or rather a variation of the old one. In 1897, Benda coined the word *mitochondria* for granules which he found first in rat spermatocytes and later in other types of cells. These mitochondria often showed a tendency to unite in the form of threads. Benda believed that his complicated technic was specific for the mitochondria, and that these represented a new discovery.

Actually, mitochondria are not to be distinguished from the granules of Altmann or the filaments of Flemming. It is important to note that not all the granules of the cytoplasm are to be considered as mitochondria. Nor is it always entirely clear just which granules are mitochondria and which ones are not. In general mitochondria are coloured with the vital dye Janus green, and Sorokin regards Janus green B as "virtually specific" for mitochondria. They can be demonstrated by the use of those fixatives which do not dissolve lipids, and it is believed that chemically mitochondria consist of a mixture of protein with a phospholipid (e.g., lecithin).

Mitochondria can be seen in living protoplasm, and they are believed to be necessary constituents of all plant and animal cells. In plant cells mitochondria may become transformed into plastids. In animal cells they have been claimed to transform into secretion granules, fibrils, and various other structures of differentiated cells. Concerning their physiological function little has been known until recently. Now it is believed that mitochondria contain oxidizing enzymes, especially the enzymes known as succinoxidase and cytochrome oxidase. Soon after the discovery of mitochondria, or rather soon after the introduction of the concept by Benda, another type of cytoplasmic structure was introduced to the cytological world.

Studying the nerve cells of an owl, Golgi described the presence of a network, the so-called "Golgi apparatus." This may be in the form of a net, or it may be in the form of granules or rods. It is demonstrated either by a silver impregnation process or by long continued soaking in osmic acid (OsO_4) solutions. For many years the Golgi apparatus was believed to have a great significance, for it could be demonstrated in many diverse types of protoplasm; and also, according to a number of authors, it could be seen in various types of living cells. Some workers have found that the region of the cell which contains the Golgi apparatus is rich in phosphatase enzymes, that is, enzymes which are able to split phosphate from organic combination.

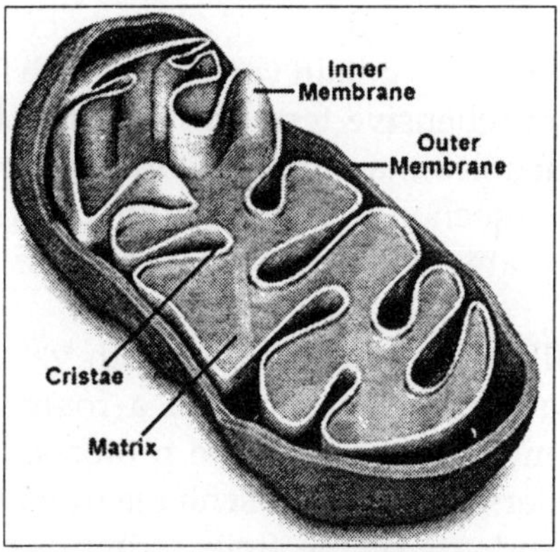

Fig. Mitochondria

However, in recent years there has been growing scepticism concerning the x. According to Hibbard, a leading worker in the field, "there does not seem to be much agreement as to the nature and functioning of the Golgi apparatus" and she feels that in so far as most of the generalizations that have been made are concerned, "we should be better off if the discovery of the impregnation methods had never been made."

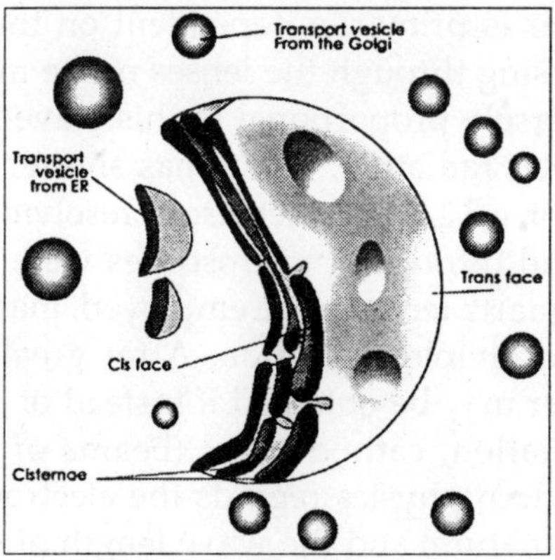

Fig. Golgi Apparatus

From time to time, it has been urged that the Golgi apparatus as such does not exist in living cells at all, but is an artifact produced by the reagents which are used to demonstrate it. Palade and Claude have presented forcible arguments in support of this view, but these arguments have been vigourously disputed.

Finally, many cells contain vacuoles, which may be minute or large. A comprehensive term for the vacuoles of a cell is "vacuome." Vacuoles may appear and disappear. They can originate from mitochondria, or be quite distinct from them. Sometimes they are produced by reagents, as will be noted later.

Occasionally cells are so filled with vacuoles that the protoplasm has the appearance of a foam. This type of protoplasm is most common in the protozoa, and it is easy therefore to understand why Bütschli regarded all protoplasm as foamlike. Students of the finer structures of cells have always attempted to use microscopes which show the greatest possible detail. For this purpose, the limiting factor is not the extent to which the microscopic image may be enlarged, for the degree of possible enlargement is infinite.

The limiting factor is the "resolving power," or the power of discriminating between 2 minute objects that lie close together. Years ago, the resolving power of ordinary microscopes was developed to its very greatest extent, for the resolving power is primarily dependent on the wave length of the light passing through the lenses of the microscope and is actually inversely proportional to this wave length. By the use of ultraviolet radiation, which has shorter wave lengths than visible light, a 2 1/2 fold increase in resolving power could be obtained, and ultraviolet microscopes were developed.

In these, quartz lenses were employed, inasmuch as glass is opaque to ultraviolet radiation. A far greater increase in resolving power may be obtained if instead of visible light or ultraviolet radiation, cathode rays (beams of electrons) are employed. Modern physics regards the electron as a particle having a wave nature, and the wave length of such radiation may be only 1/100,000th that of visible light.

Hence it might be possible to obtain vastly greater resolving power than that possible with the ordinary microscope. Developments in the new science of electron optics have made possible the construction of electron microscopes. In such microscopes the electrons are focused by electron lenses, which consist of electrostatic or magnetic fields capable of deflecting the electrons.

Cathode rays are effective only in a vacuum, and biological material to be studied in an electron microscope must be thoroughly dried. The drying process itself can cause violent distortion. Thus if some cells are suspended in a droplet of water and this is then allowed to evaporate, the surface of the water eventually encounters the cells. Surface tension exerts relatively enormous forces on small objects and the cells may be badly crushed.

To avoid such surface effects, Anderson has found it desirable to replace the water with liquid carbonic acid which is gaseous at ordinary temperatures, and then to evaporate at temperatures above the critical temperature of the gas. In this way, better pictures can be obtained.

Of course the electrons or cathode rays themselves are destructive so that living material even if it were to survive the drying would be killed by the cathode rays. In spite of all these shortcomings, the electron microscope has been able to demonstrate within bacteria structures not visible with the ordinary microscope, and it is extremely useful in the study of bacteriophages and viruses.

An electron microscope picture of bacteria and bacteriophages is shown in figure. In order to examine larger cells under the electron microscope they must either be pulled apart, which can cause violent change, or they must be studied in thin sections.

For the preparation of very thin sections, special microtomes have been constructed. So far the study of large cells with the electron microscope has not yielded results of any great interest or importance. The morphological study of the cytoplasm has resulted in a large body of information concerning the various granular and other formed constituents of living substance.

Unfortunately, however, there are no very definite clues as to the relation of these formed elements to the vital activity.

One point seems certain: the granules or filaments of the cytoplasm are not present in any spatial arrangement of significance for the physiology of the cell.

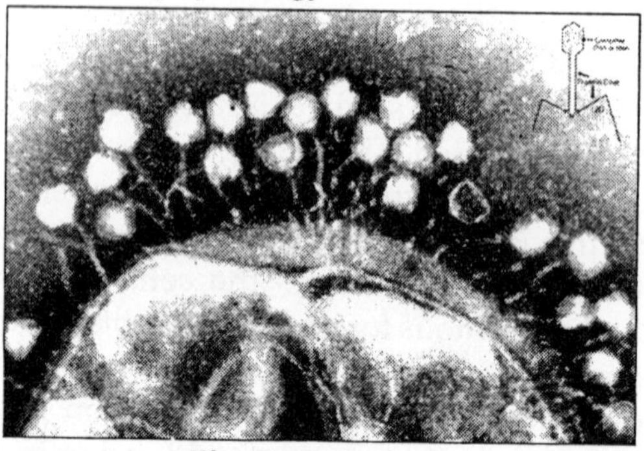

Fig. Bacteriophage

At least this is true for the main mass of the protoplasm, although it may not be true for the cortex.

When cells are centrifuged, even at very high speeds, the distribution of the granules may be profoundly altered without affecting the life or subsequent behaviour of the cell.

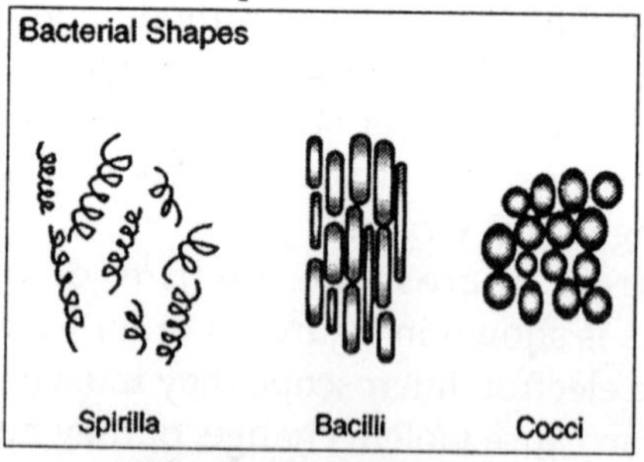

Fig. Bacterial Shapes

It should be noted that the granules of living cells may be present in tremendous numbers. In the egg of the sea urchin Arbacia, the granules occupy nearly 20 per cent of the total volume. In the egg of the starfish Asterias there is a much higher concentration of granules, and the egg of the sea urchin Paracentrotus is so full of granules that there is hardly any free space between them.

Paramecium protoplasm is also densely packed with granules. The presence of so many granules must have an important effect on the physical properties of the protoplasm as a whole. This point will be considered in a later chapter.

CHEMICAL CONSTITUENTS OF LIVING ORGANISMS

More than is often realised, scientific progress is dependent on the coining of apt phrases and catchwords. The study of granules in cells was certainly fostered by Benda's introduction of the word "mitochondria." So too the introduction of the word "vitamin" was a factor in the development of a field which had never prospered as long as vitamins were called by such awkward terms as "accessory articles of the diet."

In the study of living material the concept of a cell theory or a cell doctrine has attracted the attention of students and workers to the fact that even the most diverse types of animals and plants are constituted of the same type of unit. It is probably true that living material is as basically similar in chemical composition as in microscopical appearance; and yet this important fact, unheralded by any catchword, has never received the emphasis it deserves. True, the concept of protoplasm has had a wide influence, but the term "protoplasm" is commonly used more in a morphological sense than in a chemical one.

At the present time, indeed, one is not quite sure as to just how constant protoplasm is in a chemical sense. Numerous analyses have been made of the organs, tissues, and cells of living organisms, but these studies have been made by biochemists who have generally had little or no interest in the protoplasm concept, and were not concerned with the degree of uniformity shown by diverse types of living material.

Just as the cytologist has attempted to explain life by a study of the minute details of microscopic anatomy of cells, so the chemicallyminded biologist might seek to explain life by studying the chemical composition of living substance. As a matter of fact, however, students of the chemistry of living material have never been very optimistic in this direction. Obviously, almost all types of chemical analysis first lead to death of the cell, so that with ordinary chemical methods one can only obtain information regarding that which *was* once alive. Indeed, the chemical properties of living protoplasm are but little understood. We do not clearly understand what it is that distinguishes living protoplasm from dead, although it is certain that chemical changes do accompany and follow the death process.

Chemically, it is possible to isolate from living material various types of organic compounds. Usually, it is fair to assume that the compounds isolated were present as such when the protoplasm was alive. And yet there is good reason to believe that frequently substances obtained from cells were really in life parts of more complex compounds. So, for example, a carbohydrate extracted from living tissue may before the extraction have been bound to a protein; similarly fats or fat-like substances gotten from cells may in the living cells have been tied to protein.

Also, individual proteins which are obtained in extracts of tissues may originally have been combined with other proteins. This is now well known to be true in the case of muscle. Thus it appears that all the organic compounds known to constitute the mass of the living protoplasm may to a large extent be tied together. Also they may be tied to inorganic constituents such as phosphoric acid, sulfate, metals, alkaline earth cations or even water.

One of the most interesting aspects of the chemistry of living material is the ability of the chemist to extract from it certain catalytic substances or enzymes. These enzymes can be isolated in rather pure form and their effect on chemical reactions can be followed in test tubes. But even in the case of these enzymes, it is not certain that the extracted enzymes are in the same state as they were in the cell. Thus von Euler and von Euler urge that within the cell enzymes may exist in the form of complexes. Following extraction, the complexes are broken up into simpler constituents.

If so many substances of a cell are tied together, we might be inclined to think that protoplasm consisted in reality of giant molecules containing practically everything of importance to the life of the cell. Such ideas have been held in the past. For example, Verwor n conceived of a giant protoplasmic molecule, which he called biogen.

Verworn's ideas of the nature of biogen could hardly be held now, and indeed the modern biochemist prefers to think of protoplasm as composed of many organic compounds existing separately within the cell. If, however, at some future time viruses come to be regarded as truly living, we might have to extend our idea of protoplasm so as to include materials consisting of just one type of chemical compound.

ELEMENTS IN PROTOPLASM

There may be uncertainty as to exactly which compounds exist within the living protoplasm, or as to how the various compounds are tied together, but it is a relatively simple matter to determine exactly which elements are present and in what amount. The ordinary methods that the chemist uses for qualitative and quantitative analysis can be employed. In addition, spectographic methods are very helpful, and they can be used for minute amounts of material.

In the newer spectrography, quantitative determination of elements has been made possible by the use of photoelectric devices. The radiation emitted by an ignited substance is split up into its constituent wave lengths and the spectrum may then be recorded on photographic film or plate. Then the intensities of the spectral lines are determined by means of a densitometer or microphotometer and these intensities are compared with a calibration curve obtained from known substances in known amounts. Or without photographic film one can compare the intensity of spectral emission lines with the intensity of the lines given by known quantities of known elements.

The flame photometer is now widely used in physiological laboratories and is especially suitable for the rapid determination of potassium and sodium. In this apparatus, a solution containing sodium, potassium or other elements is fed into a flame and the emission spectrum is then compared with that of a solution containing a known amount of the element. However, it must not be supposed that the determination of elements within living material is always easy. There are difficulties which are often not fully realised. True, standard methods of quantitative analysis are applicable, but such methods can be properly used only if the material to be examined is in true solution. Usually, moreover, inorganic elements to be recognized must be in ionic form.

It is not always an easy matter to dissolve living substance completely and to free it from carbon and from other substances which would interfere with the reactions of quantitative analysis. In order to proceed with such an analysis, the living tissue must first be ashed. Because of the fact that elements like chlorine and potassium volatilize at not too high temperatures, most biochemists use what are called wet-ashing procedures. The tissue is heated with concentrated nitric or sulfuric acid or both, and in this way the carbon it contains is oxidized.

Sometimes to complete the oxidation, after ordinary acid treatment the material is dried and treated with perchloric acid. Eventually it should dissolve completely but sometimes it does not. Or even if it does, the element being tested for may still be combined with organic materials and as a result improper values are obtained when standard methods of analysis are applied. Thus, for example, when muscle is tested for chloride by the commonly used Van Slyke method, the muscle is dissolved by heating it in concentrated nitric acid, and at the same time the chloride is precipitated by silver nitrate.

Titration for chloride is then done by the standard Volhard method. Actually, this procedure, though it has been widely used, gives incorrect values for chloride. If the muscle tissue is really ashed by ignition in a tightly sealed container such as a Parr bomb, values for chloride can be obtained which are approximately twice those that can be obtained by the Van Slyke method. Even with the spectrographic methods, difficulties in ashing may be encountered. In the flame photometer the material is used in solution. A small sample of the solution is then ignited.

But in order to obtain a representative sample, it is necessary to have either a clear solution or an even suspension. Sometimes in preparing material for analysis, a few relatively large flakes of precipitate may appear. This can be very annoying, and can prevent proper analysis. Similar difficulties may arise in ordinary spectrography. Thus, in interpreting the results reported in the literature, it is often necessary to view these cautiously. In the discussion that follows, no attempt will be made at completeness.

Over 34 elements are commonly found in plant and animal tissues. Of these, carbon, hydrogen, oxygen and nitrogen are constituents of such organic compounds as proteins, fats and carbohydrates; also, to a less extent, sulfur and phosphorus.

Other elements may likewise be combined with the proteins of the cell, but ordinarily they do not form an intimate part of the protein molecule. In addition to the elements already mentioned, the following are among the elements found in practically all types of living substance: potassium, sodium, calcium, magnesium and chlorine. However, many plants can grow in the apparent absence of sodium and, indeed, some of the common nutrient solutions used for the growth of plants contain no sodium.

Lithium is typically present in marine animals, but according to Fox and Ramage, lithium, although widespread, is not a universal constituent of protoplasm. Rubidium is also rather widespread in its occurrence. It was found in all human tissues examined except bone, being especially abundant in heart and striated muscle. Although higher plants seem to require calcium, some fungi are said to grow normally without it. It is by no means an easy matter to remove traces of sodium or calcium from nutrient solutions, so that it is difficult to determine whether or not living material can exist in the total absence of these elements.

Strontium is close to calcium chemically and appears to be able to replace it in some biological systems. A small amount of strontium is generally present in marine invertebrates . Magnesium is a constituent of chlorophyll, and is therefore found in all green plants. Although sea water itself contains several times as much magnesium as calcium, the animals which live in sea water commonly contain less magnesium than calcium. This is not always true, however, and some gastropods including the very common marine snail Littorina littorea contain more magnesium than calcium. Some insects are also richer in magnesium than in calcium. Rats cannot live without a small amount of magnesium.

If deprived of it, the calcium-magnesium balance is upset, and convulsions and death follow. Boron appears to be an essential constituent of citrus plants, tobacco, and tomato. It is also found in small amounts in Marine Invertebrates. Aluminum is a very common constituent of living material, especially in plants. It is also found in animal tissues. Heavy metals, although not present in large amounts, are very important for life. In organic combination they act as carriers of oxygen (in the blood) and as oxidizing catalysts within cells. Especially important are iron and copper.

Iron is a necessary constituent of the blood cells of all vertebrates and of many invertebrates. As Warburg first showed, cellular respiration is primarily dependent on the presence of iron-containing catalysts believed to be present in all living cells.

The blood cells of many invertebrates contain copper and the blue colour of crab and limulus blood is due to the presence of this element. Lindow, Elvehjem, and Peterson have shown a widespread distribution of copper in living tissues; Sheldon and Ramage conclude that it is a universal constituent of living protoplasm.

This opinion is shared by Marston. The literature on the copper content of living systems is now very large, mainly because of the importance of copper in cell metabolism. Zinc is widely distributed among both plants and animals. Galtsoff found it in many marine invertebrates. It has also been reported for the silkworm and for vertebrates generally. Gettler and Bastian describe a simple method of detecting zinc in tissues and they find rather high amounts (50 mg. per kilo) in human liver and kidney.

Keilin and Mann showed that zinc is a constituent of the important enzyme carbonic anhydrase and Smirnov finds that the zinc content of the red blood cells of various vertebrates runs parallel to their carbonic anhydrase content. Zinc is also present in blood plasma.

The retina, iris, and choroid of vertebrate eyes and especially the eyes of fish are rich in zinc and this is also to be correlated with a high carbonic anhydrase content of these tissues. In view of the fact that the insulin molecule is thought to contain zinc, it is not surprising that the pancreas should be rich in this element. Manganese is also an important element.

It is essential for the growth of rats and chickens, and also for the growth of various types of plants. Webb found manganese generally present in invertebrates; according to Vinogradoff, ants have an especially high content. Human blood contains only a small amount of it. In almost all plant and animal tissues, it is possible to detect cobalt and nickel by means of extremely delicate tests. The quantities found range from 0.002 to 2 mg. per kilo.

All types of animals that have been tested contain molybdenum; tunicates are especially rich in it. Marston (loc. cit). believes that molybdenum is probably always present in living matter. Some metallic elements are of rather sporadic occurrence. Small amounts of arsenic and of silicon have been found in various tissues of man and higher animals. Ohlmeyer and Olpp found quite appreciable amounts of silicon in the blood, lung and pancreas of the calf. It had previously been found in connective tissue, hair and feathers.

Some plants seem to have a definite need for silicon. Webb found tin in the starfish, the sea urchin, and in a gastropod mollusk. There is a literature on the presence of vanadium in the blood cells of ascidians. The element is also present in other tissues of ascidians but not in detectible amounts in rat tissues. Vanadium occurs in a wide variety of plants. Traces of uranium have been found in the fungi Aspergillus niger and Ustilago carbo; also in human kidney, brain, thyroid and blood.

The rare element gallium is apparently essential for Aspergillus. Many human organs contain mercury in small amounts. In some parts of the world gold is found in the soil and is taken up by plants. Deer eating such plants accumulate gold in their horns, hoofs and hair. There is a large body of evidence to show that metallic elements present in traces have a profound effect on the life of various types of plants. The subject is of great practical importance. Of the non-metallic elements, as already mentioned, chlorine is probably found in all types of living material. For many years, it was believed that chlorine although present in tissues generally, did not occur within the cells but only in the spaces between the cells.

This doctrine was at one time held for various types of vertebrate tissues and organs, but especially for muscle. Analysis of muscle by the standard Van Slyke method showed relatively low concentrations of chlorine, and this could all be accounted for if it were assumed that the tissue spaces of the muscle contained chlorine in the same concentration as that in which it occurs in the blood. As a matter of fact, as already noted, the Van Slyke method gives improper values for the chloride of muscle. Moreover, chloride in reasonably high concentrations can actually be demonstrated in isolated frog muscle fibers.

Fluorine is of little importance for animals and plants, for with few exceptions it occurs only in minute amounts and it is not known to have any well-defined function. The same applies to bromine. As is commonly known, iodine is an important constituent of the thyroid gland of vertebrates. Iodine is also abundant in marine brown algae and is obtained commercially from this source.

COMPOUNDS OF PROTOPLASM

Although it is simpler to determine the chemical *elements* present in living substances, one can also attempt an analysis of the chemical *compounds* which are present. It is not difficult to determine the percentage of various classes of substances. Numerous such analyses have been made for tissues, organs, and whole animals. Analyses of cells are less common. The following table gives the percentage composition of the egg of the sea urchin, Paracentrotus lividus, and is taken from Ephrussi.

	Per cent
Water	77.3
Protein	15.18
Fat and lipoid	4.81
Carbohydrate	1.36
Ash	0.34
	98.99

This may be compared with an analysis of the human body shown in the following table.

	Per cent
Water	66.0
Protein	16.0
Fat and lipoid	13.0
Carbohydrates	0.61
Ash	5.0

In the above analysis the figures for fat and ash are high, doubtless because the body as a whole contains many non-protoplasmic materials. As compared with animals, plants are relatively rich in carbohydrates. Analyses of plant tissues show a content of carbohydrate material which may be equal to or even in excess of the protein content. *Water.*

As already indicated, water is the chief chemical constituent of protoplasm. Concerning it Adolph says: "The vortex of vital activities is, with respect to water, more than analogous to the whirlpool. Without water both the body and the pool would be but paleontological specimens, showing only (in Huxley's phrase) where activity has been." Most of the determinations of the water content of living materials have been made by drying these materials in an oven at temperatures slightly above the boiling point of water.

The loss of weight is assumed to represent the original water content. This procedure doubtless involves some loss of volatile matter other than water. Also, because of the presence of humidity within the oven all the water is not driven off. Use of a vacuum oven, or of circulating hot air, can reduce the humidity error.

As a matter of fact, the two types of error tend to neutralize each other, although the error due to incomplete dryness in the oven is relatively large. Because of the humidity error, many of the older determinations of the water content of organisms and tissues may be too low. A simple and rapid method of water determination is to place tissues or small organisms in toluene or some similar liquid and distill off the water. This is then collected and its volume determined. There is also a titration method for water determination. Some organisms have very high percentages of water.

Thus according to Vernon, the water content of the ctenophore Cestus is 96.2 per cent; the jellyfish Rhizostoma has 96.0 per cent, the tunicate Salpa 96.3 per cent, and the nudibranch Tethys 95.3 per cent. Paramecium has 79 per cent water.

Even in the tissues of higher animals the water content may be rather high. Thus in the gray matter of the brain and in the kidney the water concentration is above 80 per cent; so that in these organs there is actually a higher concentration of water than in the fluid blood. The lowest concentrations of water are found in seeds and in certain types of animals which have the power of drying up.

Rotifers and tardigrades normally live in ponds and puddles. In dry seasons, as the ponds or puddles dry up, the rotifers lose practically all their water. In such dried-up forms protoplasmic activity ceases almost entirely. As a matter of fact, protoplasmic activity is associated with water content. Young and growing animals and plants are richer in water, and are more active than somewhat older organisms.

Most of the water of a living cell is free, that is to say, it is not bound chemically to other chemical constituents of the protoplasm. However, a certain percentage of water may be "bound."

The subject of bound water is a controversial one and at present is in an unsettled state. Proteins can bind water chemically. Certainly this is true of dried proteins or of crystalline proteins. But whether the water bound in a crystal remains bound when the crystal dissolves is of course doubtful. Bound water is defined in various ways.

For most authors bound water is water which is not available for dissolving solutes. Such water may be bound chemically or it may be held physically either by adsorption or by being trapped in the interstices of a gel or jelly. According to one author, water is free if it evaporates according to the usual kinetics of evaporation, but is bound if it is retained by the medium and evaporates less actively than free water.

Concerning the amount of bound water in colloidal solutions of proteins and other substances, there is dispute. Newton and Gortner determined percentages of bound water in plant juices by studying the effect of adding sugar on the freezing point of these juices.

Chambon and Gouttebarge found large amounts of bound water in plant gums, mucilages, and pectins, as well as in animal materials such as the vitreous humour of the eye or the mucus secreted by the foot of a snail. They believe that there is a correlation between the amount of bound water and the uronic acid content of a substance.

Table. Inorganic Composition of Longitudinal Muscles of Caudina Chilensis (The Figures Represent Equivalents Per 100 Gm. Fresh Weight)

Cations		*Anions*	
K	0.01385	Cl	0.01220
Na	0.01910	SO4	0.01305
Ca	0.01775	CO3	0.01700
Mg	0.00785		
Total	0.05855	Total	0.04225

Chambon and Gouttebarge used the Nicloux method, in which the amount of water available for the solution of alcohol is determined. In this method very dilute solutions of alcohol are used and the method is available for living systems. Nicloux himself used his method in this way and he determined the amount of bound water in small (whole) fishes and in various tissues of larger fishes and frogs. This work has perhaps not had the attention it deserves.

This analysis differs from most analyses of muscle in that it shows more sodium than potassium. Typically muscle is much richer in potassium than in sodium. It will be noted that the sum of the cation equivalent weights is decidedly greater than that of the anions.

Organic Constituents

Living substance contains three main types of organic constituents. Of these the proteins and carbohydrates are typically water-soluble substances. It is true that some proteins and some carbohydrates are not soluble in water or in dilute salt solutions, but insoluble proteins are relatively rare, and insoluble carbohydrates are easily broken down to soluble compounds.

On the other hand, the third class of substances characteristic of protoplasm is distinguished by a thorough-going insolubility in water. Indeed, all protoplasmic constituents not soluble in water, but soluble in fat solvents, are grouped together as lipids. The term "lipids" can then be taken to include, first, the true fats, secondly the phosphatids which are compounds of fatty acid, phosphoric acid, and organic base, and, thirdly, the sterols.

The latter bear no chemical kinship to the fats, but resemble them in being insoluble in water but soluble in ether. Although not so abundant in living material as are the proteins, the lipids are indispensable for life. Insoluble in water, they are nevertheless finely dispersed in the watery protoplasm.

This is in part due to the fact that the lipids are emulsified; in part, also, to the fact that the lipids may combine with proteins to form water-soluble compounds. The nature of these compounds is not at all well known, but it seems certain, as already noted, that they do occur in protoplasm.

The combination of lipid with protein is but one example of the fact that within the living cell various types of chemical compounds may be linked to each other. At the present time, our knowledge of actual combinations of organic substances in the living protoplasm is meager. What we know chiefly is that the organic substances which do constitute the protoplasm are of such a nature that they might well be assumed to combine.

Thus, if some of the proteins contained in cells behave primarily like acids and others like bases, it is only natural to assume that these proteins might be linked together in life. Indeed the existence of combined proteins is well known. Carbohydrates may be combined with proteins not only in true chemical combination but also the carbohydrates may be tied to the proteins by loose secondary linkages.

Various alkaline and alkaline earth cations can be and undoubtedly are bound to the proteins of the protoplasm, and they may also be linked to lipids or even to carbohydrates. Perhaps also anions may be bound to the proteins of cells. Indeed, although no one at the present favours the old biogen concept of protoplasm, there is a growing fund of knowledge to indicate that many of the chemical compounds isolated from dead cells may in the living organism be linked by chemical bonds.

OSMOTIC RELATIONS OF CELLS

In view of the fact that living cells are surrounded by osmotic membranes, and are thus tiny osmotic engines, it is important for the physiologist and for the general biologist to know something about osmotic phenomena and the laws of osmosis.

Indeed, in any investigation in which cells or tissues are immersed in a solution it is essential that this solution have the same osmotic properties as the fluids in which the cells and tissues are normally bathed. Similarly, if a solution is injected into the blood stream of an animal, it should not differ markedly in osmotic properties from the blood itself.

Many a biological investigation has been spoiled by a failure to observe this rule, or by an ignorance of how to arrive at a proper osmotic concentration. In the following discussion no attempt will be made at completeness, and very little attention will be paid to theories of osmosis or to mathematical treatment of osmotic laws. Moreover, although solutions in all types of solvents show osmotic behaviour, we shall confine our attention to watery solutions, for these are the only types of solutions with which we shall have to deal in living systems. When two solutions are separated by a semipermeable membrane, there is a flow of solvent toward the solution of higher molecular concentration.

This constitutes osmotic flow, and the pressure developed as a result of this flow is osmotic pressure. A semipermeable membrane is a membrane which permits solvent to pass, but not dissolved substance or solute.

Actually no perfect semipermeable membrane is known; at any rate, we know of no membrane which permits water to pass and resists the passage of all dissolved substances. However, there are membranes which are permeable to water and impermeable to certain dissolved substances, and if we choose the proper solutions, these membranes are truly semipermeable.

PERMEABILITY OF THE CELL MEMBRANE

The osmotic behaviour of a living cell depends in large measure on the characteristics of the membrane which surrounds it. As pointed out in the previous chapter, only those substances which are incapable of passing through it or which pass through with difficulty produce osmotic pressures against it. It is important, therefore, to have information concerning the permeability of the plasma membranes of cells if we are to have a true picture of the osmotic relationships. There are also other reasons for studying cell permeability: every cell is dependent on its environment.

It obtains nutritive substances from the fluids which surround it, and to these surrounding fluids it passes out excretory substances. All of this exchange takes place through the cell membrane, and the rate of exchange is of course governed by the permeability of the membrane. Moreover, this permeability is not constant but variable: it changes with the physiological state of the cell; it is influenced also by external conditions such as temperature or light or the chemical constitution of the environment.

In the changing permeability of the cell many authors have thought to find an explanation for various types of vital activity.

These authors apparently regard cell permeability as the chief variable of those which are known or knowable in the life of the protoplasm. Moreover, some authorities believe that the differences in electric potential between various parts of living systems and the changes in these potentials during vital activity are primarily due to membrane permeability differences.

Finally, a knowledge of the permeability of the osmotic membrane or plasma membrane of the cell gives information as to the essential properties of this membrane, and may even lead to conclusions as to its chemical make-up. To avoid confusion, the concept of permeability needs to be clearly defined.

According to Brooks and Brooks, it is "the rate of movement of a substance through the permeable layer under a given driving force.

This definition involves two concepts which must be kept sharply distinguished: on the one hand, permeability—a property of a membrane or region of a cell: on the other hand, the driving force which may be independent of any property of the membrane."

The driving force is primarily due to a difference in concentration of a substance on the two sides of the membrane or permeable layer. This driving force becomes progressively less as equilibrium is approached, finally reaching zero at equilibrium. "Permeability measures the rate at which equilibrium is approached, that is, what fraction of the total change necessary for the attainment of equilibrium is acomplished in unit time."

In non-living systems, ordinarily when a membrane separates two concentrations of a given substance to which the membrane is permeable, equilibrium is reached when the concentrations at both sides of the membrane are equal. But in a living cell such an equilibrium may never occur and the cell may reach a condition of no further change and yet possess either more or less of the substance than the surrounding medium. This may be due in part to adsorption (at the surface or within the cell) or to chemical reactions. In addition, it must be remembered that the cell is alive and that a portion of its vital energy may be expended in furthering the accumulation or the disposal of one substance or another.

Thus, when there is no further change in the ratio of concentration (of a given substance) inside and outside of the cell, a true equilibrium may not exist, but rather a dynamic equilibrium.

Such an equilibrium is generally referred to as a "steady state." Normally a given cell is exposed to water plus a variety of substances in solution.

The membrane of the cell has a definite, although doubtless a somewhat variable, permeability toward each of the substances to which it is exposed, as well as to any other substance which may be added. In the past confusion has arisen because authors have frequently written of permeability in a rather loose fashion.

Thus it is not sufficiently accurate to say that the permeability of a cell increases-we must know to what substance or substances the permeability is increased. Recent literature indicates that permeability to one substance may increase, whereas permeability to another may decrease.

In discussing the permeability of the cell we shall for the most part consider only the plasma membrane. Actually, when a dissolved substance penetrates a plant cell, other membranes than this are also involved.

The nonliving cell wall of plant cells is ordinarily permeable both to water and to dissolved substances, but in special cases, as for example in barley and wheat seeds, the cell wall offers resistance to the passage of dissolved substances.

Plant physiologists often study the passage of substances into the large vacuole of a plant cell; in order to reach this vacuole from the exterior of the cell, substances must pass through not only the plasma membrane but also a membrane or film which separates the vacuole from the protoplasm.

It is well to remember this in interpreting the many studies in which the penetration of substances into plant vacuoles has been measured.

Of course, it is possible that the properties of the membrane lining the vacuole are similar to those of the outer cell surface, but this has never been proved, and there is indeed some evidence to the contrary.

There have been many studies on cell permeability. The earlier investigations were concerned mostly with the qualitative information as to whether or not the cell membrane permitted the passage of this or that substance, or as to whether various conditions and reagents caused increase or decrease of permeability.

In the more recent literature there has been an increasing effort to make quantitative determinations of the speed of entrance of various substances into the cell (or out of it). We shall consider both the permeability of the cell membrane to dissolved substances and to water.

For the older literature on permeability the monograph of Stiles is very useful. More recent books on permeability are those of Brooks and Brooks and Davson and Danielli. The discussion in Höber's Physical Chemistry of Cells and Tissues may profitably be consulted. Wilbrandt has published some useful reviews.

The results obtained in the study of cell permeability are not always in agreement. This is due in part to the fact that cells of different types do not always show the same permeability relations toward dissolved substances.

Another reason for disagreement lies in the fact that the various methods are not equally dependable. Indeed, some of the indirect methods are open to serious question. It is important to examine the various methods critically, as Stiles and Jacobs have done.

According to Jacobs, "It will appear that no single method by itself is entirely reliable, and the obvious conclusion to be drawn is that in any given investigation as many and as diverse methods as possible ought to be employed before an attempt is made to base generalizations on the results obtained."

One difficulty common to all methods is the sensitivity of the plasma membrane. When cells are injured the semipermeability of the membrane is apt to be more or less completely destroyed. Care must be taken, therefore, to avoid injury as much as possible.

If, for example, one is studying the permeability to acids, it is hard to be sure that even in dilute solutions the acid does not produce some measure of injury.

Some methods depend on the exposure of the cell to hypotonic or hypertonic solutions, and such solutions may also change the permeability of the membrane.

Chapter 3

Shelter for Animals

BEEKEEPING

Fixed Frame Hives

There are considerable regional variations in the type of hive in which bees are kept. A hive is a set of rectangular wooden boxes filled with moveable wood or plastic frames, each of which holds a sheet of wax or plastic foundation. The bees build cells upon the sheets of foundation to create complete honeycombs. Foundation comes in two cell-sizes: worker foundation - which enables the bees to create small, hexagonal worker cells and 'drone foundation' which allows the bees to build much larger cells - drone cells, for the production of male bees. The bottom box, or brood chamber, contains the queen and most of the bees; the upper boxes, or supers, contain just honey.

Only the young nurse bees can produce wax flakes which they secrete from between their abdominal plates; they build honeycomb using the artificial wax foundation as a starting point, after which they may raise brood or deposit honey and pollen in the cells of the comb. These frames can be freely manipulated and honey supers with frames full of honey can be taken and extracted for their honey crop.

MODERN BEEKEEPING

Movable Frame Hives

In the USA, the Langstroth hive is commonly used. The Langstroth was the first successful top-opened hive with movable frames, and other designs of hive have been based on it. Langstroth hive was however a descendant of Jan Dzierzon's Polish hive designs. In the United Kingdom, the most common type of hive is the British National Hive, but it is not unusual to see some other sorts of hive.

Straw skeps, bee gums, and unframed box hives are now unlawful in most US states, as the comb and brood cannot be inspected for diseases. However, straw skeps are still used for collecting swarms by hobbyists in the UK, before moving them into standard hives.

Top Bar Hives

A few hobby beekeepers are adopting various top bar hives of the type commonly found in Africa. These have no frames and the honey filled comb is not returned to the hive after extraction, as it is in the Langstroth hive. Because of this, the production of honey in a top bar hive is only about 20 per cent that of a Langstroth hive, but the initial costs and equipment requirements are far lower.

Top-bar hives also offer some advantages in interacting with the bees and the amount of weight that must be lifted is greatly reduced. Top Bar Hives are being widely used in developing countries in Africa and Asia as a result of 'Bees For Development' programme.

Protective Clothing

Beekeepers often wear protective clothing to protect themselves from stings. While knowledge of the bees is the first line of defence, most beekeepers also wear some protective clothing.

Novice beekeepers usually wear gloves and a hooded suit or hat and veil. Experienced beekeepers sometimes elect not to use gloves because they inhibit delicate manipulations. The face and neck are the most important areas to protect, so most beekeepers will at least wear a veil.

Defensive bees are attracted to the breath, and a sting on the face can lead to much more pain and swelling than a sting elsewhere, while a sting on a bare hand can usually be quickly removed by fingernail scrape to reduce the amount of venom injected.

The protective clothing is generally light coloured and of a smooth material. This provides the maximum differentiation from the colony's natural predators (bears, skunks, etc.). which tend to be dark-coloured and furry.

AQUACULTURE

Aquaculture is the cultivation of aquatic organisms. Unlike fishing, aquaculture, also known as aquafarming, implies the cultivation of aquatic populations under controlled conditions. Mariculture refers to aquaculture practiced in marine environments. Particular kinds of aquaculture include algaculture (the production of kelp/seaweed and other algae); fish farming; shrimp farming, shellfish farming, and the growing of cultured pearls.

Aquaculture has been used in China since circa 2500 BC. When the waters lowered after river floods, some fishes, namely carp, were held in artificial lakes. Their brood were later fed using nymphs and silkworm feces, while the fish themselves were eaten as a source of protein. By a fortunate genetic mutation, this early domestication of carp lead to the development of goldfish in the Tang Dynasty.

The Hawaiian people practiced aquaculture by constructing fish ponds. A remarkable example from ancient Hawaii is the construction of a fish pond, dating from at least 1,000 years ago, at Alekoko. According to legend, it was constructed by the mythical Menehune. The Japanese practiced cultivation of seaweed by providing bamboo poles and, later, nets and oyster shells to serve as anchoring surfaces for spores. The Romans often bred fish in ponds.

The practice of aquaculture gained prevalence in Europe during the Middle Ages, since fish were scarce and thus expensive. However, improvements in transportation during the 19th century made fish easily available and inexpensive, even in inland areas, causing a decline in the practice.

The first North American fish hatchery was constructed on Dildo Island, Newfoundland Canada in 1889, it was the largest and most advanced in the world. Americans were rarely involved in aquaculture until the late 20th century, but California residents harvested wild kelp and made legal efforts to manage the supply starting circa 1900, later even producing it as a wartime resource.

In contrast to agriculture, the rise of aquaculture is a contemporary phenomenon. According to professor Carlos M. Duarte About 430 (97 per cent) of the aquatic species presently in culture have been domesticated since the start of the 20th century, and an estimated 106 aquatic species have been domesticated over the past decade. The domestication of an aquatic species typically involves about a decade of scientific research.

Current success in the domestication of aquatic species results from the 20th century rise of knowledge on the basic biology of aquatic species and the lessons learned from past success and failure. The stagnation in the world's fisheries and overexploitation of 20 to 30 per cent of marine fish species have provided additional impetus to domesticate marine species, just as overexploitation of land animals provided the impetus for the early domestication of land species

In the 1960s, the price of fish began to climb, as wild fish capture rates peaked and the human population continued to rise. Today, commercial aquaculture exists on an unprecedented, huge scale. In the 1980s, open-netcage salmon farming also expanded; this particular type of aquaculture technology remains a minor part of the production of farmed finfish worldwide, but possible negative impacts on wild stocks, which have come into question since the late 1990s, have caused it to become a major cause of controversy.

TYPES OF AQUACULTURE

Algaculture

Algaculture is a form of aquaculture involving the farming of species of algae. The majority of algae that are intentionally cultivated fall into the category of microalgae, also referred to as phytoplankton, microphytes, or planktonic algae.

Macroalgae, commonly know as seaweed, also have many commercial and industrial uses, but due to their size and the specific requirements of the environment in which they need to grow, they do not lend themselves as readily to cultivation on a large scale as microalgae and are most often harvested wild from the ocean.

Fish Farming

Fish farming is the principal form of aquaculture, while other methods may fall under mariculture. It involves raising fish commercially in tanks or enclosures, usually for food. A facility that releases juvenile fish into the wild for recreational fishing or to supplement a species' natural numbers is generally referred to as a fish hatchery.

Fish species raised by fish farms include salmon, catfish, tilapia, cod, carp, trout and others. Increasing demands on wild fisheries by commercial fishing operations have caused widespread overfishing. Fish farming offers an alternative solution to the increasing market demand for fish and fish protein.

Freshwater Prawn Farming

A freshwater prawn farm is an aquaculture business designed to raise and produce freshwater prawn or shrimp for human consumption. Freshwater prawn farming shares many characteristics with, and many of the same problems as, marine shrimp farming.

Unique problems are introduced by the developmental life cycle of the main species. The global annual production of freshwater prawns (excluding crayfish and crabs) in 2003 was about 280,000 tons, of which China produced some 180,000 tons, followed by India and Thailand with some 35,000 tons each. Additionally, China produced about 370,000 tons of Chinese river crab (*Eriocheir sinensis*).

Integrated Multi-trophic Aquaculture

Integrated Multi-Trophic Aquaculture (IMTA) is a practice in which the by-products (wastes) from one species are recycled to become inputs (fertilizers, food) for another. Fed aquaculture is combined with inorganic extractive (e.g. seaweed) and organic extractive (e.g. shellfish) aquaculture to create balanced systems for environmental sustainability (biomitigation), economic stability (product diversification and risk reduction) and social acceptability (better management practices).

"Multi-Trophic" refers to the incorporation of species from different trophic or nutritional levels in the same system. This is one potential distinction from the age-old practice of aquatic polyculture, which could simply be the co-culture of different fish species from the same trophic level. In this case, these organisms may all share the same biological and chemical processes, with few synergistic benefits, which could potentially lead to significant shifts in the ecosystem.

Some traditional polyculture systems may, in fact, incorporate a greater diversity of species, occupying several niches, as extensive cultures (low intensity, low management) within the same pond. The "Integrated" in IMTA refers to the more intensive cultivation of the different species in proximity of each other, connected by nutrient and energy transfer through water, but not necessarily right at the same location.

Ideally, the biological and chemical processes in an IMTA system should balance. This is achieved through the appropriate selection and proportions of different species providing different ecosystem functions.

The co-cultured species should be more than just biofilters; they should also be harvestable crops of commercial value. A working IMTA system should result in greater production for the overall system, based on mutual benefits to the co-cultured species and improved ecosystem health, even if the individual production of some of the species is lower compared to what could be reached in monoculture practices over a short term period.

Sometimes the more general term "Integrated Aquaculture" is used to describe the integration of monocultures through water transfer between organisms. For all intents and purposes however, the terms "IMTA" and "integrated aquaculture" differ primarily in their degree of descriptiveness. These terms are sometimes interchanged. Aquaponics, fractionated aquaculture, IAAS (integrated agriculture-aquaculture systems), IPUAS (integrated peri-urban-aquaculture systems), and IFAS (integrated fisheries-aquaculture systems) may also be considered variations of the IMTA concept.

Mariculture

Mariculture is a specialized branch of aquaculture involving the cultivation of marine organisms for food and other products in the open ocean, an enclosed section of the ocean, or in tanks, ponds or raceways which are filled with seawater. An example of the latter is the farming of marine fish, prawns, or oysters in saltwater ponds. Non-food products produced by mariculture include: fish meal, nutrient agar, jewelries (e.g. cultured pearls), and cosmetics.

Shrimp Farming

A shrimp farm is an aquaculture business for the cultivation of marine shrimp for human consumption. Commercial shrimp farming began in the 1970s, and production grew steeply, particularly to match the market demands of the U.S., Japan and Western Europe. The total global production of farmed shrimp reached more than 1.6 million tonnes in 2003, representing a value of nearly 9,000 million U.S. dollars. About 75 per cent of farmed shrimp is produced in Asia, in particular in China and Thailand. The other 25 per cent is produced mainly in Latin America, where Brazil is the largest producer. The largest exporting nation is Thailand.

Shrimp farming has changed from traditional, small-scale businesses in Southeast Asia into a global industry. Technological advances have led to growing shrimp at ever higher densities, and broodstock is shipped world-wide. Virtually all farmed shrimp are penaeids (i.e., shrimp of the family *Penaeidae*), and just two species of shrimp—the *Penaeus vannamei* (Pacific white shrimp) and the *Penaeus monodon* (giant tiger prawn)—account for roughly 80 per cent of all farmed shrimp.

These industrial monocultures are very susceptible to diseases, which have caused several regional wipe-outs of farm shrimp populations.

Increasing ecological problems, repeated disease outbreaks, and pressure and criticism from both NGOs and consumer countries led to changes in the industry in the late 1990s and generally stronger regulation by governments. In 1999, a programme aimed at developing and promoting more sustainable farming practices was initiated, including governmental bodies, industry representatives, and environmental organizations.

Housing Buffalow

Housing for water buffaloes should give protection against thermal stress - particularly direct exposure to sun, heavy rains and cold weather. It must allow good ventilation. Housing may therefore be different in different areas of the world, due to differences in climate.

Below are some considerations and solutions when planning housing in hot and cold climates respectively. Common for all housing is that enough space should be allowed for each buffalo.

The outdoor yard should preferably be covered with grass or maybe concrete, in order to prevent it from becoming an unhygienic mud hole in rainy periods.

Hot Climate

Buffaloes may appear to be misplaced in a hot and humid environment. They have a dark skin and few sweat glands and are more or less dependent on water for their cooling. This is not entirely true, buffaloes protected from direct sunlight do very well even during hot and humid days, partly because their ability to loose heat through the respiratory tract.

High milk production requires a high feed intake which leads to higher metabolic heat production. High yielding buffaloes thus have a disadvantage over lower yielding animals, and need more cooling facilities. The following points are guidelines to have in mind when giving advice on management.

- The feeding, watering and milking place should always give shade and protection from heavy rains, either by trees or by a roof.
- Cool water either from a clean river or served in an earthen pit, helps the animals to maintain temperature. Water trough should always be placed in the shade.
- A paddock with trees gives a very cheap and effective protection from sun. However, the trees may need to be protected from the buffaloes also.
- A shelter of a simple construction with only a roof. In hot humid climates it is better not to have walls. Walls may lead to inadequate ventilation and thereby favouring bacteria and mold growth, thus making the stable unhygienic. To protect the inside from sunshine (or heavy rain), curtains made from straw, textile or other suitable material, can be used.
- Providing the animals with a wallow. However, the wallow should be one with clean water and not far from the farm. Spending time walking in the sun to and from the wallowing costs more than it saves.
- Showering of the buffaloes with cool water for 3 minutes twice a day has proven to be an efficient way for them to get rid of excess heat.

Cold Climate

- A shelter should protect the animals from rain, snow and strong wind. It may be a simple construction with a roof and three walls. This system will allow the buffaloes to go outside to graze when the weather allows it. There should be a feeding area inside the shelter in case of several days with bad weather. A separate heated milking area is advisable in this case.
- A dry and clean bedding is important in cold weathers to maintain animal health.

- In case of extremely cold climate, (Caucasia and Balkan) with several months with a temperature below 0?C, a heated barn may be necessary.

Pens for Calves

Calves should be kept in individual pens for the first month. The pens should be easy to keep clean, with shelter from direct sunlight, rain, snow and draught. By keeping the calves individually it is easier to check that they eat and grow properly and to detect illnesses. Also, naval suckling is avoided and spread of diseases is more difficult.

The calves should have access to fresh and clean water at all times. Preferably, the buckets for milk and water should be outside the pen, in a steady holder within easy reach for the calf. Hereby, the calves can not splash it on the bedding. A humid bedding will facilitate growth of germs and parasites. The pen should contain a holder for hay and concentrate. These holders should be placed above the ground so that the calf cannot step or defecate in them.

Chicken Coop

A chicken coop is a building where chickens are kept on a farm or homestead.

Fig. Chicken Coop

The inside of the chicken coop normally consists of several nesting boxes just large enough for the chickens to sit in while laying their eggs, as well as perches for the chickens to use while sleeping. Often the floor of the chicken coop is covered with a material, such as straw or wood chips, to contain the chicken waste, and to allow easier cleanup of the chicken coop.

Sometimes the chicken coop contains feeding and watering devices for the chickens, but these may also be kept outside of the chicken coop. Most have some kind of ventilation to help air out any odors that may occur.

Since chickens are often prone to illness from drafts and poor weather, chicken coops are best designed to be warm at cold times, cool at hot times, and as draft free as possible. Many people, especially in rural areas, keep a small flock of chickens to provide themselves with eggs and meat.

These small coops are often surrounded by a fence, usually made with wire fencing to allow the chickens an area to roam, peck, and hunt insects, to keep them contained. Contrary to popular use of the term, fencing known as "chicken wire" is often too weak to pose a barrier to predators and is therefore seldom used to confine the birds.

Free Range

Free range poultry farming consists of poultry permitted to roam freely instead of being contained in any manner. The principle is to allow the animals as much freedom as possible, to live out their instinctual behaviours in a reasonably natural way, regardless of whether or not they are eventually killed for meat. In the UK, the Department for Environment, Food and Rural Affairs says that a free range chicken must have daytime access to open-air runs during at least half of their life. Unlike in the United States, this definition also applies to eggs. The European Union regulates marketing standards for egg farming which specifies a minimum condition for Free Range Eggs states that "hens have continuous daytime access to open-air runs, except in the case of temporary restrictions imposed by veterinary authorities".

NEST BOX

A nest box also called birdhouse is a man-made box provided for animals to nest in. Nest boxes are most frequently utilized for wild and domesticated birds, but certain species of mammal may also use them. Boxes are usually wooden, though some for birds are made from mixtures of wood and concrete, called *woodcrete.*

Metal nest boxes are also marketed, but these are generally unsuitable for outdoor use, as they can overheat easily in sunshine. The majority are cuboid, and many have a sloping roof. Many have a hinged top to provide access for cleaning, bird ringing or, when used for domesticated species, to give the breeder access to the young. Boxes may either have an entrance hole or be open fronted. Some nest boxes can be highly decorated and complex. Designs mimicking human house design or other structures are common in some gardens.

The size of a nest box should be suited for the guest inside it. Quite large boxes are used for ducks and owls. Nest boxes for wild birds should be cleaned annually to remove old nests and kill parasites.

The nest boxes of domestic birds and other animals should be removed and cleaned after the breeding season is over and no more young are expected. Budgerigars and other birds raised domestically will not usually mate unless a nest box is present. Bats houses are becoming popular sources of natural mosquito and insect control. A single bat can eat 500 to 1,000 mosquitoes a night as well as numerous garden pests.

Directions for making the open bottom bat houses for small and large colonies, as well as locations to purchase them are available on the internet.

Bat houses are an ecologically friendly way of controlling mosquitoes as pesticides mostly kill the mosquito's predators rather than mosquitoes. Nest boxes are marketed not only for birds but also for bats, some other mammals and bees. Depending on the animal, these boxes are used for roosting, breeding, or both. Most small mammals will use a nest box for raising young if one is provided.

Kennel

A kennel is the name given to any structure or shelter for dogs. A *kennel* is a doghouse, run, or other small structure in which a dog is kept. American English often uses the word "doghouse" in this context; in British English the term "kennel" is preferred, but the phrase "in the doghouse" is understood but does not mean "in the kennel", rather it is a colloquial phrase meaning (of a person) "in trouble" or "being shunned".

Used in the plural, *the kennels,* the term means any building, collection of buildings or a property in which dogs are housed, maintained, and bred.

Stable

A stable is a building in which livestock, especially horses, are kept. It most commonly means a building that is divided into separate stalls for individual animals. There are many different types of stables in use today such as the American barn which is a large barn with a door each end and individual stalls inside or free standing stables with the classic top and bottom opening doors.

Historically, the headquarters of a unit of cavalry rather than just their horses' accommodation, would be called a stable.Used metaphorically from this origin, a stable is a collection of people (e.g. professional wrestlers) working under a single manager or animals kept by one owner.

Yarding

While often confused with free range farming, yarding is actually a separate method by which a hutch and fenced off area outside are combined when farming poultry. The distinction is that free-range poultry are either totally unfenced, or the fence is so distant that it has little influence on their freedom of movement. This is common technique used by small farms in the Northeastern portion of the US. Daily releases out of hutches or coops allows for instinctuial nature for the chickens with protections from predators. The hens usually lay eggs either on the ground of the coop or in baskets if provided by the farmer.

This technique can be complicated if used with roosters though, mostly because of difficulty getting them into the coop and to clean the coop while it is inside. This territorial nature is apparent while outside in which they have a brood of hens and sometimes even informal land claims. This can endanger people unaware of the existence of the territories who are attacked by the larger birds.

Housing goats

Housing for dairy goats does not have to be elaborate, but it must satisfy the health and comfort of the animals. There are five requirements of good housing for goats:

- The building should be adequately ventilated but not drafty;
- The walls and ceiling should be free from condensation;
- The bedded area should be relatively dry and clean;
- The hay, grain and water receptacles must be well built and located so that feed is not wasted or contaminated; and
- The facilities should provide easy access to the animals and require a minimum amount of labour.

The comfort zone for dairy goats is between 55 and 70 degrees Fahrenheit. Non-sweating animals are much less sensitive to declining temperatures than to rising temperatures. Milk production, feed consumption and comfort are not affected by temperatures between 0 and 55 degrees Fahrenheit, but temperatures over 80 degrees Fahrenheit seriously reduce feed intake and milk output. Therefore, the object is not how to keep the goats warm in winter, but how to keep them cool in summer.

The movement of air, either by mechanical or natural means, to remove heat, moisture and odors is a necessary part of a housing plan. Most pneumonia problems with dairy goats can be traced to inadequate ventilation. Wet walls and ceilings are the result of improper ventilation, poor insulation or a combination of the two.

The rate of air movement is influenced by the amount of animal heat produced and the temperature you wish to maintain in the building. Additional heat and insulation may be required to keep the stable air fresh and to prevent water pipes from freezing in the winter. An air inlet system also must be provided for good air distribution. Ventilation entails more than installing a fan to move some air.

Proper ventilation during the summer may require moving 150 to 200 cubic feet of air per minute per animal. Winter weather may reduce the amount of air to be exhausted to as little as 20 cubic feet per goat. Contact your local MU Extension specialist for assistance in planning the right ventilation system.

With proper ventilation, you are protecting the animals, the building and the quality of milk produced. Windows are essential in a closed barn. They permit sunlight for warmth and drying and provide a source of vitamin D for the animals. Well-lighted barns usually are kept cleaner. In summer, open windows are important for air movement.

Housing for Young Stock and the Milking Herd

There are two housing systems often used for goats:

- Loose housing, where the animals run loose in a pen or shed
- Stall barns, where each animal is confined in a small box stall or tie stall

Loose housing is an old system with many desirable features. With plenty of bedding, the manure pack for a goat herd can be kept fairly clean and dry. Heat produced within the pack makes a warm bed. There is ample exercise area and an opportunity for the goats to move around.

This type of housing, however, presents a few problems. A built-up manure pack, even though dry on the surface, releases a lot of moisture into the barn air. In addition, hay racks should be located off the bedded area to reduce parasite infestations. Watering devices should be placed in an area where spilled water will not mess up the bedded pack. Many goat owners use a shed or barn that is open on the south or southeast side, thus eliminating the need for mechanical ventilation.

These are cold housing facilities that do not require additional heat or insulation and use normal air movement for ventilation. Goats move freely in or out of the housing area and into the paddock or feeding area. Hay feeders, watering devices, mineral feeders and grain bunks are located on concrete pads some distance from the built-up manure pack.

Housing for the kids and other young stock may be included in the plan for the milking herd. However, young goats must be kept separate from the milking herd. Frequently, the kids are kept in 4-foot-square box stalls with at least one side slatted to permit air movement. Do not make a box stall from plywood with all four walls solid; air movement will not reach the kids.

Usually a gated front contains enough openings to allow air drainage. Provide a heat lamp for newborn kids and for kids that are ill. Older kids may be kept in a corner of the stable area inside a pen made from movable hurdles. Buck housing must be separate and downwind from the milking herd. It does not have to be elaborate and often is no more than a 5- or 6-foot shed with an open side facing the south to give the bucks free access to an exercise lot.

The milking area is part of the housing plan, but it should be separated from the stable area. It should have a concrete floor to make cleaning easier, and the milking platform should be 15 to 18 inches higher than the floor to permit easier milking. The platform should allow 18 inches in width and 3-1/2 feet in length for each animal to be tied. The manger should be 6 inches deep and 1 foot wide. If several goats are to be fastened and milked at once, a lever that will open and close the stanchion head locks is helpful.

Some goat owners who keep their milking herds in stall barns milk their animals on a milking stand located in one corner of the stable. Dusty air and flies in the stable, however, may require you to place the milking stand in a separate screened-in facility to guarantee clean milk. A 5-by-8-foot room with a concrete floor and drain is adequate for milking a small herd, but be sure there is plenty of light either from natural sources or electric bulbs.

If the milk is sold to the public or to a processor, there will be state inspection of the operation just as is required for production of Grade A milk from cows. Check with your local health authorities and the state inspector before starting construction.

If the milk is to be used by the family, the degree of sanitation and cleanliness maintained and the size and type of building constructed can be based on the owner's good sense, pride, conscience and finances. Equip the milk house with a double sink for washing utensils, a hot water heater, refrigeration, a small table or work space, and a rack for drying and storing utensils. At least 50 percent of the milk house should be open floor space with ample room for equipment.

Regardless of the herd size, refrigeration must be available to quickly cool the milk. A household refrigerator may be used, but cold water is more efficient in cooling milk than cold air. The refrigerator or cooler should be large enough to hold a pan of water. The pan should hold enough water to match the amount of milk in the milk bucket. If the herd is big enough and milk production is 5, 10 or more gallons per day, it will be necessary to purchase a water-immersion cooler or a bulk tank for cooling purposes.

Cooling is critical to milk flavor and quality. All milk contains bacteria, some of which comes from the air and the utensils. If milk remains warm for a short period, the bacteria begin to multiply and the quality of milk deteriorates. Therefore, cool milk immediately after milking to a temperature under 40 degrees Fahrenheit and hold it at that temperature until processed and/or consumed. Plans for building a milking barn and milking stand and for construction of the milk house area are available from your local MU Extension dairy specialist.

Fences serve two purposes that make them an important part of a housing plan — they keep goats in and dogs out. They also keep foraging goats away from your trees and shrubs as well as your neighbour's flower patch. Set 7-foot posts, either wooden or steel, on 12-foot centres.

Then use either a 4-foot-high woven wire fence topped with an electric wire 12 inches above the woven wire or use a completely electric fence. If you use a completely electric fence, place the first wire 12 inches from the ground, then four more wires on 6-inch spacings to a total height of 36 inches. Use insulators on the post to attach the wires.

The electric fence works well when weeds, grass and brush are not allowed to touch the wires and short it out or when the ground isn't bone dry during periods of drought. Your local farm supply store can recommend the proper equipment.

In summary, keep the plans as simple and economical as possible. Consider the health of your animals as well as the conveniences a housing facility might offer. Contact your local MU Extension specialist for additional ideas on housing facilities for your dairy goat enterprise.

Chapter 4

Nutrition for Animals

FEEDING GOATS

Feeds for goats can be divided into several different categories: forages, grains, protein supplements, minerals and water. Depending on what levels of each category that is fed. It will be found that goats will gain about.25 to.5 a pound each day.

Goats will readily eat a variety of forages. These include pasture, hay, haylage, and brush. Any species of forage is acceptable for goats, although be careful of feeding a lot of alfalfa to wethers and billies because the higher calcium levels can lead to urinary calculi or kidney stones. Goats are very useful in many operations for cleaning up unwanted brushy areas.

Most brushy areas will only be able to survive about 2 years with intensive defoliation. Forages are an important part of a ruminant's ration because the high fibre is necessary to keep the digestive system healthy. Grains provide the energy portion of a ration for goats. A variety of grains are available, with corn, oats, wheat and barley being the most widely used. Limit the amount of wheat in the ration to no more than 2 per cent to prevent any digestive problems. When starting goats on a feedlot ration, be sure to gradually introduce them to grain.

Feeding high levels of grain suddenly can cause bloating and overeating disease. Disease can be prevented by vaccinating with type C and D perfringens. This is often available in one vaccine, along with tetanus. Any male goats that are castrated should receive a dose of tetanus vaccine. Castration is a personal consideration as many ethnic groups prefer to purchase intact males.

Protein Supplements

Market goats should be fed a grain ration that runs between 14 per cent and 18 per cent protein. To reach these protein levels, we need to include some sort of protein supplement. Typical supplements will include soybean meal, roasted soybeans, cottonseed, brewers or distiller's grains, or flour mill by-products. Beware of using any dairy protein supplements because of the amount of copper in the supplements. Many nutritionists still disagree over the amount of copper required in a goat ration, but care should still be given to overfeeding this nutrient.

Minerals

One of the best ways to provide minerals to a market goat is through a commercially mixed salt and mineral mix. Be sure to purchase a mix formulated specifically for goats. Allow goat's free access to a mineral feeder that contains the mineral mix at all times. Minerals are especially important for helping to ensure that goats remain healthy. These mineral mixes will often include some of the essential vitamins for goats as well.

Water

Water is often one of the most overlooked aspects of livestock feeding operations. Clean, fresh water should be available at all times for all species of livestock. Animals that have adequate amounts of water are more likely to stay healthy as well as have a faster growth rate. Goats will often be able to supply all of their water needs through dew and lush pasture. However, it is still in their best interest to have a supply of water available to them at all times.

OTHER FEED ADDITIVES

For herds that are prone to urinary calculi, ammonium chloride can be added to prevent the disease. Another common practice is to add a coccidiostat to the ration to prevent as well as control coccidiosis. Common additives used to prevent coccidiosis include Rumensin, Bovatec or Deccox (decoquinate). Lastly, no matter what the ration, be sure to provide enough feeder space for all the goats to eat. Plan for at least 1 foot of feeder space for each goat. And, don't forget to clean feeders and waterers on a regular basis.

Sound feeding practices and good housing facilities result in optimum growth and high milk production and contribute to the good health and comfort of goats.

When feeding goats, keep these objectives must be in mind:

- Feed a young animal enough energy for growth, and feed a mature animal enough energy to maintain a fairly constant body weight;
- Provide enough protein, minerals and vitamins in a balanced feeding programme to maintain a healthy animal; and
- Offer does enough extra food during gestation and lactation for fetus development and milk production.

Optimum growth, good health and high milk production are the results of sound feeding practices. Dairy goats are not unique in their body requirements; they will respond to good nutritional practices. Digestible fibre is especially important in dairy goat diets. Too much grain in relation to forage does not foster good ruminant action and is a costly feeding practice. When feeding, keep minerals and trace mineral salt separate. Always feed them separately.

Feed hay in a rack that will not permit wasting. One recommended type of feeder is a keyhole feeder. Do not overlook forage testing as an economical way to feed the correct ration to herd.

Feeding Kids and Yearlings

Kids are born with no natural protection from disease. The first milk (colostrum) from the mother offers protection and gets the digestive system working. To be most effective, it must be fed before disease-producing organisms enter the mouth and digestive tract. Wash the fresh doe's udders and teats with warm water.

Hand milk half a cup of colostrum and feed it to the kid within 15 minutes of birth. This is the best way to ensure that the newborn receives some milk and to provide it with the most protection from organisms present on the skin of the doe. Complete the first milking and store the colostrum for later feedings if hand feed is elected. Otherwise, permit the kid to nurse at its convenience following the first hand feeding.

Clean the feeding utensils immediately after each use. The same cleaning procedures should be followed for washing milk-handling equipment. Milk replacer may be fed from the fourth day; it should contain at least 20 percent protein, 20 percent fat and be free of vegetable products. A lamb or high-quality calf milk replacer is recommended. Provide hay and grain at one to two weeks of age.

Wean from milk when grain intake reaches 1/4-pound daily and kids are readily consuming hay. Make sure the solution is thoroughly mixed. Mix a fresh mixture daily and feed in place of milk. Feed as soon as diarrhea is noticed. Use for 1-1/2 or 2 days, then return to the regular milk diet. After four to six months of age, the kids may be fed a ration similar to that of the milking herd. Good hay and 1/2 pound of grain per day should provide an ample growth rate. Poor hay may require 1 to 1-1/2 pounds of grain daily.

Feeding the Milking Herd

If milk production is important, feed maximum amounts of high quality hay balanced with a grain ration containing enough protein, minerals and vitamins to support production and animal health. Grass or legume hays are equally acceptable. As the percentage of legumes is increased, the need for protein in the grain mix is reduced.

To determine the amount of grain to feed, consider level of milk production, amount and quality of forages consumed, appetite and state of fleshing. Thin, high-producing does should have access to all the hay they can eat plus grain to the limit of their appetite. Does in mid-lactation that are in good flesh should have all the hay they will eat plus 1 pound of grain for each 3 pounds milk produced. Late lactation does may not need more than 1 pound of grain for each 5 pounds of milk.

Feed a grain ration formulated for a milk-producing ruminant (dairy cows). Rolled or cracked grain is more palatable than ground grain. Because of palatability problems, urea is not recommended. Some commercial cow feeds may contain byproduct ingredients unpalatable to goats. Wet molasses is more palatable than dry molasses. Beer or citrus pulp is a valuable source of fibre, especially if the available hay is of low quality.

If the doe is not thin, reduce the amount of grain to 1/2 to 1 pound per day. Feed her all the forage she will eat. Hay fed during the dry period may be of lower quality, but if so, the grain ration should contain additional protein. Browse, leaves and weeds are often useful to recondition the stomach. If the dry ration differs from the milking ration, be sure to change to the milking forage and grain ration two weeks before the doe freshens. Most bucks do not need more than a pound of grain per day plus forages. Don't let them grow fat. Adjust grain upward or downward accordingly. Always feed full forage.

FEEDING AND NUTRITION OF PIG

Feeding guinea pig is a fairly simple manner, though there are a few special needs that we should be aware of. Guinea pigs are strict vegetarians, and in addition to their pellets, they will eat fruits and vegetables, bread, and will even nibble on vegetarian dog biscuits.

Before getting in to the details, it is important to point out that guinea pigs like a routine, predictable life. Sudden changes in their diet can cause stress, so be careful. If the brands of pellets is changed, gradually by adding a few of the new ones to their old pellets, and slowly switching the mixture over. Feeding excessive amounts of fruits and vegetables at one time can cause diarrhea, particularly if new foods are introduced.

All guinea pigs need water to survive. Though they can obtain a large portion of their water from fresh fruits, vegetables and greens, they will still require water from a water bottle to stay healthy. This water should be changed daily, and the water bottles themselves washed periodically to keep them clean. To prevent algae growth, make sure that the sun does not directly shine on the water bottle for long periods of time.

If there is arrangement of tap water bottled water or a water filter system should be considered. Distilled water is not recommended for regular drinking, as it can disturb the osmotic pressure in the body cells, which can potentially lead to osmotic shock.

Be aware that some guinea pigs like to "play" with their water bottle, leaking out large amounts of water by holding the steel ball up in the tube. Others like to try and blow water back up into the bottle, which can contaminate the entire supply.

Hay is the mainstay of every guinea pig's diet. Though guinea pig pellets do contain hay, it is not in sufficient quantity to keep guinea pig healthy. Guinea pig must have a fresh supply of hay every day in order to keep its digestive system regular, and as a general rule.

Timothy hay is the best type of hay to feed, though any grass hay will do. If grass hay is not available, a legume hay can be fed, such as alfalfa, though in general this should be avoided as much as possible. Legume hays are high in calcium, and this can lead to bladder stones (uroliths) in some animals.

Most pet stores carry dried timothy hay in large and small bags, but it is far better to purchase hay fresh from a feed and garden store, or directly from a farmer. Buying hay by the bale is especially economical, as a bale of hay will literally last for months, and will stay fresh as long as it is stored properly. Some tips on hay storage can be found here.

Guinea Pig Pellets

Guinea pig pellets will provide guinea pig with the proper balance of vitamins (save for Vitamin C), minerals and other nutrients. Although it's not necessary to feed pellets to guinea pigs, it is certainly recommended that they be primary feed, after hay. Guinea pig pellets are, in particular, a prime source of protein; obtaining the necessary amounts of this and other nutrients will require careful dietary planning if pellets are not chosen.

As a general rule of thumb, adult guinea pigs will eat between one and two ounces of pellets each day. More active animals will eat less, and animals without much stimulation will eat more out of boredom. Nursing and pregnant sows may demand a little more than this, and animals that are given plenty of hay and some fresh vegetables each day might eat a little less. It will be easy to tell if they are fed too much or too little: if there are leftover pellets. If the bowl is empty and they seem to be foraging for food, then they aren't fed enough.

Try to feed at the same hour each day, as guinea pigs like a routine and predictable life. Remove any leftover pellets after an hour or so. Guinea pig pellets are a prime source of protein: a diet that is heavy on pellets will lead to a fat guinea pig. Hay is the only food that they should have access to 24 hours a day.

When choosing pellets, be sure to *avoid* the guinea pig "mixes" that contain nuts, seeds and dried fruits. These mixes are high in fats and oils, which can lead to excessive weight gain. Additionally, many of these mixes contain sunflower seeds in their shells. Guinea pigs should *never* be fed nuts or seeds that are still in their shells (peanuts, sunflower seeds, etc.): dozens of guinea pigs in the Pacific Northwest alone die each year from choking on these shell fragments.

One thing that should be noted before we move on: guinea pigs should be fed guinea pig pellets, not rabbit pellets or pellets for another species of animal. Guinea pig pellets are nutritionally balanced for guinea pigs, and feeding them pellets for other animals can result in serious health problems.

Though some people may tell that their guinea pigs survive well on rabbit pellets, for instance, it is not trusted that all rabbit food is safe for guinea pigs. Their particular brand may work for guinea pigs, but another brand may lack necessary vitamins or minerals, or have them in insufficient or excessive quantities.

There is no regulation in the pet food industry, so pellets and feeds vary considerably between manufacturers. Don't play Russian roulette with cavies: only feed them pellets that are explicitly and *exclusively* for guinea pigs.

It is *extremely important* that guinea pigs receive enough vitamin C each day to prevent scurvy. Guinea pigs cannot manufacture or store vitamin C, so they must obtain it from their diet on a daily basis. Adult guinea pigs will require 10mg of vitamin C each day, and nursing and pregnant sows will require twice that amount (20mg). The caveat: vitamin C, also known as ascorbic acid, breaks down very quickly, so it must be supplied fresh. Although many pellet manufacturers will claim that their feed contains the required amount of vitamin C.

Vitamin C in pellets breaks down during storage, and after about 90 days, there won't be enough vitamin C in the feed to keep guinea pigs healthy. Also note that the food starts breaking down on the shelf, not when the bag of pellets are opened. Unless the manufacturer dates the bag, showing when the feed was packaged, we really have no idea how long pellets have been on the shelf at the local pet store.

Since pellets are generally an inadequate source of vitamin C, most owners will either feed sufficient quantities of fruits and vegetables rich in vitamin C each day, or provide vitamin C in the form of supplements added to the guinea pigs' food or drinking water.

Vitamin supplements come in many forms: pre-packaged vitamin supplements for guinea pigs can be bought from pet store, or self grind up vitamin C tablets. In either case, most people prefer to add the vitamins to the animal's drinking water, as that is the only way to guarantee that it will be ingested daily. If the supplement routes are chosen, here are a few pointers:

- Add the appropriate amount of supplement to the amount of water that the guinea pig will drink in a day. Most water bottles work best if they are filled halfway or more, so end up wasting a lot of vitamins. Fortunately, they aren't very expensive. If guinea pig drinks 4 fluid ounces of water a day, and fill water bottle with 8 ounces, then need to add 20mg of vitamin C supplement. This will lead to 4 ounces of wasted water and 10 mg of wasted vitamin C.

- Vitamin C (ascorbic acid) breaks down very quickly in water, so need to add it on a daily basis.
- The chlorine in tap water can actually inactivate ascorbic acid. If vitamin supplements in water is opted, then use tap water that has been standing for at least 24 hours, or filtered tap water from a filter system that reduces chlorine content.

Being vegetarians, guinea pigs will eat many kinds of fruits, vegetables and fresh greens. And, many fruits and vegetables are sources of vitamin C, providing a natural way to meet guinea pig's daily requirement.

Calcium and Phosphorus totals for the given food amounts are presented as well. Although these are necessary minerals, and a part of every guinea pig's diet, they are getting them from their pellets, as well. The nutrients guinea pig is getting along with their vitamin C. Too much calcium can cause bladder sludge, a precursor to bladder stones. The proper balance of phosphorus, calcium and vitamin C prevents guinea pigs from needing vitamin D supplements.

Although most guinea pig pellets contain vitamin D as a precaution, there's no guarantee that particular brand of feed does. In addition to providing vitamin C and variance in cavy's diet, fresh greens and fruits contain a significant amount of water. Feeding "wet foods" will aid nursing sows in lactating, and will reduce the amount of water that the average guinea pig will drink from the water bottle.

It is especially useful to take lettuce, green peppers and other watery greens on trips to provide moisture for guinea pigs while in the car, since water bottles will leak heavily as the bumps in the road rattle the ball in the drinking tube.

To reduce or completely eliminate pellets from guinea pig's diet, they can survive on water, hay and fresh fruits and vegetables. However, this will require very careful dietary planning on part, to make sure that they receive the necessary nutrients in the proper amounts. Also, be aware that many fresh greens are laxative in action, which means that run the risk of giving guinea pigs loose bowels, or even diarrhea, if the are fed too much at one time. If runny droppings are noticed immediately cut fresh greens out of their diet and feed dry foods until the feces returns to normal. Vegetables that are not laxative (such as carrots) may still be fed.

Guinea pigs will not eat what they don't like: some cavies have very discriminating tastes, while others will eat anything that put in front of them. Each pig has its own preferences, and it may take some time to figure out what he or she likes best. Given below is a list of foods that guinea pigs can/will eat. This list is neither comprehensive nor complete.

Apples, bananas, bread (slightly stale & crunchy, but not moldy), broccoli, carrot greens, carrots and baby carrots, celery (cut into small pieces first), cilantro, cucumber, dandelion greens, grass, green & red bell peppers, green leaf & romaine lettuce, kale, kiwi, mustard greens, oats, oranges, parsley, raspberries, spinach, tomatoes.

When feeding "wild" greens, such as grass and dandelion greens, make sure they have not been sprayed with chemicals, or contaminated by droppings or urine from other animals, such as cats, dogs and birds. The dangers of pesticides are obvious, and feces can carry any number of parasites which can be transmitted to cavies eating contaminated greens. Even if dogs and cats are dewormed regularly, their feces can contain protozoa/bacteria which can cause debilitating diarrhea in a guinea pig.

Some foods to avoid are listed below:

- Long celery stalks
- Iceberg lettuce
- Any shelled nuts or seeds
- Raw beans
- Rhubarb

There is some confusion as to whether or not potato peelings are good or bad: some books indicate that they are poisonous to guinea pigs, while others say that they are okay in small amounts. The truth is that potatoes are *okay*, however, any *green* in a potato *is poisoinous*.

If potato peelings are given, make sure there are *no green spots* anywhere in the portions. Also, some people may recommend yogurt in small amounts, for it's bacteria-growing properties (to aid the digestive system). Instead, look into acidophilus powders or liquids: acidophilus is also a bacteria growing culture, and it';s likely to give better results since less of it needs to be fed.

There are several commercially available treats on the market that are aimed at guinea pigs. Berry flavoured "crunchies" and flavoured chew sticks are popular among some owners and guinea pigs, and are generally safe. Some owners also have experimented with vegetarian dog biscuits and dry cereals such as Cheerio's.

However, the guinea pig "treat sticks", which are also available commercially. These sticks are essentially seeds and nuts that are held together with honey. Most of the treat sticks on the market contain sunflower seeds that are still in their shells, which is a big no-no for guinea pigs due to the choking hazards.

FEEDING BEHAVIOUR OF BUFFALOES

Buffaloes are strict grazers and only browse when feed is utterly scarce. Normally, buffaloes graze during the day. In case of extremely high ambient temperatures, grazing takes place in the morning and afternoon and sometimes during night time. Buffaloes graze more and better than cattle. Thereby they consume more feed and nutrients per kg of body weight than cattle do.

Newborn calves suckle their mothers within two hours of birth. Normal suckling frequency is approximately 6 to 8 times per day. The calves start to nibble grass at 3 to 4 weeks of age, although they are not actually grazing until after a few weeks more. When the calves have reached two months of age, forage starts to become more important than previously and soon most of the nutrient intake comes from forage rather than milk. Natural weaning of calves is usually within a year or before its mother's next parturition.

NUTRITION FOR HORSE

Equine nutrition refers to the feeding of horses, ponies, mules, donkeys, and other equines. Correct and balanced nutrition is a critical component of proper horse care. Horses are herbivores, a type of non-ruminant known as a "hind-gut fermentor." What this means is that horses have only one stomach, similar to humans. However, unlike humans, they also have to digest plant fibre that comes from grass and hay. Therefore, unlike ruminants, who digest fibre in plant matter by use of a multichambered stomach, horses use bacterial fermentation that occurs in the organ known as the *cecum* (or *caecum*) to break down cellulose.

In practical terms, horses prefer to eat small amounts of food steadily throughout the day, as they do in nature when grazing on pasture. Although this is not always possible with modern stabling practices and human schedules that favour feeding horses twice a day, it is important to remember the underlying biology of the animal when determining what to feed, how often, and in what quantities. The digestive system of the horse is somewhat delicate.

Because horses are unable to regurgitate food, except from the esophagus, if they overeat or eat something poisonous, vomiting is not an option. They also have a long, complex large intestine and a balance of beneficial bacteria in their cecum that can be upset by rapid changes in feed. Because of these factors, they are very susceptible to colic, which is a leading cause of death in horses. Therefore, horses require clean, high-quality feed, provided at regular intervals, and may become ill if subjected to abrupt changes in their diets.

Horses are also sensitive to molds and toxins. For this reason, they must never be fed contaminated fermentable materials such as lawn clippings. Fermented silage, sometimes called "haylage" is fed to horses in some places; however, contamination or failure of the fermentation process that allows any mold or spoilage may be toxic. Horses and other members of the *Equidae* family are adapted by evolutionary biology to eating small amounts of the same kind of food all day long.

In the wild, horses ate prairie grasses in semi-arid regions and traveled significant distances each day in order to obtain adequate nutrition. Therefore, their digestive system was made to work best with a small but steady flow of food that does not change much from day to day. Digestion begins in the mouth. First, the animal selects pieces of forage and picks up finer foods, such as grain, with sensitive, prehensile, lips.

The front teeth of the horse, called incisors, clip forage, and food is ground up for swallowing by the premolars and molars. The esophagus carries food to the stomach. The esophagus enters the stomach at an acute angle, creating a one-way valve, with a powerful spincter mechanism at the gastroesophageal junction, which is why horses cannot vomit. The esophagus is also the area of the digestive tract where horses may suffer from choke.

Horses have a relatively small stomach for their size, which limits the amount of feed that can be taken in at one time. The average sized horse has a stomach with a capacity of only four gallons, and works best when it contains about two gallons. One reason continuous foraging or several small feedings per day are better than one or two large meals is because the stomach begins to empty when it is 2/3 full, whether the food in the stomach is processed or not.

The small intestine is 50 to 70 feet long and holds 10 to 12 gallons. This is the major digestive organ where most nutrients are absorbed. Bile from the liver acts here, combined with enzymes from the pancreas and small intestine itself. Equids do not have a gall bladder, so bile flows constantly, another reason for a slow but steady supply of food.

Most nutrition is absorbed into the bloodstream from the small intestine. The cecum is the first section of the large intestine. It is also known as the "water gut" or "hind gut." It is a cul-de-sac pouch, about 4 feet long that holds 7 to 8 gallons. It digests cellulose plant fibre through bacterial fermentation. The reason horses must have their diets changed slowly is so the bacteria in the cecum are able to modify and adapt to the different chemical structure of new feedstuffs. Too abrupt a change in diet can cause colic, as the new food is not properly digested.

The large colon, small colon, and rectum make up the remainder of the large intestine. The large colon is 10-12 feet long and holds up to 20 gallons of semi-liquid matter. Its main purpose it to absorb carbohydrates which were broken down from cellulose in the cecum.

Due to its many twists and turns, it is a common place for a type of horse colic called an impaction. The small colon is 10-12 feet long, holds about 5 gallons, is the area where the majority of water is absorbed, and where fecal balls are formed. The rectum is about one foot long, and acts as a holding chamber for waste, which is then expelled from the body via the anus.

Nutrients

Like all animals, equines require five main classes of nutrients to survive: water, energy (primarily in the form of fats and carbohydrates), proteins, vitamins, and minerals.

Water makes up between 62-68 per cent of a horse's body weight and is essential for life. Horses can only live a few days without water, becoming dangerously dehydrated if they lose 8-10 per cent of their natural body water. Therefore, it is critically important to provide access to a fresh, clean, and adequate supply of water. An average 450kg horse drinks 10 to 12 gallons of water per day, more in hot weather, when eating dry forage such as hay, or when consuming high levels of salt, potassium, and magnesium.

Horses drink less water in cool weather or when on lush pasture, which has a relatively higher water content. When under hard work, or if a mare is lactating, water requirements may be as much as four times greater than normal. Though they need a great deal of water, horses spend very little time drinking; usually 1-8 minutes a day, spread out in 2-8 episodes. Water plays an important part in digestion. The forages and grains horses eat are mixed with saliva in the mouth to make a moist bolus that can be easily swallowed. Therefore, horses produce up to 10 gallons of saliva per day.

Nutritional sources of energy are fat and carbohydrates. Protein is a critical building block for muscles and other tissues. Horses that are heavily exercised, growing, pregnant or lactating need increased energy and protein in their diet. However, if a horse has too much energy in its diet and not enough exercise, it can become too high-spirited and difficult to handle.Fat exists in low levels in plants and can be added to increase the energy density of the diet.

Fat has 9 Mcal/kg of energy, which is 2.25 times that of any carbohydrate source. Because equids have no gall bladder to store large quantities of bile, which flows continuously from the liver directly into the small intestine, fat, though a necessary nutrient, is difficult for them to digest and utilize in large quantities. However, they are able digest a greater amount of fat than can cattle. Horses benefit from up to 8 per cent fat in their diets, but more does not always provide a visible benefit. Horses can only have 15 per cent-20 per cent fat in their diet without the risk of developing diarrhea.

Carbohydrates, the main energy source in most rations, are usually fed in the form of hay, grass and grain. Soluble carbohydrates such as starches and sugars are readily broken down to glucose in the small intestine and absorbed. Insoluble carbohydrates, such as fibre (cellulose), are not digested by enzymes, but are fermented by microbes in the cecum and large colon to break down and release their energy sources, the volatile fatty acids.

Soluble carbohydrates are found in nearly every feed source; corn has the highest amount, then barley and oats. Forages normally have only 6 to 8 per cent soluble carbohydrate, but under certain conditions can have up to 30 per cent. Sudden ingestion of large amounts of starch or high sugar feeds can cause colic or laminitis.Protein is used in all parts of the body, especially muscle, blood, hormones, hooves, and hair cells. The main building blocks of protein are amino acids. Alfalfa and other legumes in hay are good sources of protein that can be easily added to the diet. Most adult horses only require 8 to 10 per cent protein in their diet; however, higher protein is important for lactating mares and young growing foals.

Horses that are not subjected to hard work or extreme conditions usually have more than adequate amounts of vitamins in their diet if they are receiving fresh, green, leafy forages. Sometimes a vitamin supplement is needed when feeding low-quality hay, if a horse is under stress (illness, traveling, showing, racing, and so on), or not eating well. Grain has a different balance of nutrients than forage, and so requires specialized supplementation to prevent an imbalance of vitamins and minerals.

Minerals are required for maintenance and function of the skeleton, nerves and muscles. These include calcium, phosphorus, sodium, potassium, and chloride, and are commonly found in most good-quality feeds. Horses also need trace minerals such as magnesium, selenium, copper, zinc and iodine. Normally, if adult animals at maintenance levels are consuming fresh hay or are on pasture, they will receive adequate amounts of minerals in their diet, with the exception of sodium chloride (salt), which needs to be provided, preferably free choice. Some pastures are deficient in certain minerals, especially selenium or copper, and in such situations, deficiency diseases may occur if the horses' feed is not supplemented.

Calcium and phosphorus are needed in a specific ratio of between 1:1 and 2:1. Adult horses can tolerate up to a 5:1 ratio, foals no more than 3:1. A total ration with a higher ratio of phosphorus than calcium is to be avoided. Over time, the imbalance will ultimately lead to bone problems such as osteoporosis.

Foals and young growing horses through their first three to four years have special nutritional needs and require feeds that are balanced with a proper calcium:phosphorus ratio and other trace minerals. A number of skeletal problems may occur in young animals with an unbalanced diet. Hard work increases the need for minerals; sweating depletes sodium, potassium, and chloride from the horse's system. Therefore, supplementation with electrolytes may be required for horses in intense training, especially in hot weather.

Types of Feed

Equids can consume approximately 2 per cent to 2.5 per cent of their body weight in dry feed each day. Therefore, a 1000 lb adult horse could eat up to 25 pounds of food. Foals less than six months of age eat 2 to 4 per cent of their weight each day.Solid feeds are placed into three categories: forages, concentrates (including grain or pelleted rations), and supplements.

Equine nutritionists recommend that 50 per cent or more of the animal's diet by weight should be forages. If a horse is working hard and requires more energy, the use of grain is increased and the percentage of forage decreased so that the horse obtains the energy content it needs for the work it is performing. However, forage amount should never go below 1 per cent of the horse's body weight per day.

Forages, also known as "roughage," are plant materials classified as legumes or grasses, found in pastures or in hay. Often, pastures and hayfields will contain a bland of both grasses and legumes. Nutrients available in forage vary greatly with maturity of the grasses, fertilization, management, and environmental conditions. Grasses are tolerant of a wide range of conditions and contain most necessary nutrients. Some commonly used grasses include timothy, brome, fescue, coastal bermuda, orchard grass, and Kentucky bluegrass.

Legumes such as clover or alfalfa are usually higher in protein, calcium, and energy than grasses. However, they require warm weather and good soil to produce the best nutrients. Legume hays are generally higher in protein than the grass hays.

They are also higher in minerals, particularly calcium, but have an incorrect ratio of calcium to phosphorus. Because they are high in protein, they are very desirable for growing horses or those subjected to very hard work, but the calcium:phosphorus ratio must be balanced by other feeds to prevent bone abnormalities.

Hay is a dried mixture of grasses and legumes. It cut in the field and then dried and baled for storage. Hay is most nutritious when it is cut early on, before the seed heads are fully mature and before the stems of the plants become tough and thick. Hay that is very green can be a good indicator of the amount of nutrients in the hay; however, colour should not be used as sole indicator - smell and texture are also important. Hay can be analyzed by many laboratories and that is the most reliable way to tell the nutritional values it contains. Moldy or dusty hay should not be fed to horses, as it is the most common cause of Recurrent airway obstruction, also known as COPD or "heaves."

Hay, particularly alfalfa, is sometimes compressed into pellets or cubes. Processed hay can be of more consistent quality and is more convenient to ship and to store. It is also easily obtained in areas that may be suffering localized hay shortages. However, these more concentrated forms can be overfed and horses are somewhat more prone to choke on them. On the other hand, hay pellets and cubes can be soaked until they break apart into a pulp or thick slurry, and in this state are a very useful source of food for horses with tooth problems such as dental disease, tooth loss due to age, or structural anomalies.

Haylage, also known as *Round bale silage* is a term for grass sealed in airtight plastic bags, a form of forage that is frequently fed in the United Kingdom and continental Europe, but is not often seen in the United States. Because haylage is a type of silage, hay stored in this fashion must remain completely sealed in plastic, as any holes or tears can stop the preservation properties of fermentation and lead to mold or spoilage. Rodents chewing through the plastic can also spoil the hay introducing contamination to the bale.If a rodent dies inside the plastic, the subsequent botulism toxins released can contaminate the entire bale.

Another type of forage sometimes provided to horses is beet pulp, a byproduct left over from the processing of sugar beets, which is high in energy as well as fibre.Sometimes, straw or chaff from cereal grains is fed to animals. However, this is roughage with little nutritional value. It is not recommended by nutritionists as a horse feed, though it is sometimes used as a filler; it can slow down horses who eat their grain too fast, or it can provide additional fibre when the horse must meet most nutritional needs via concentrated feeds. Straw is more often used as a bedding in stalls to absorb wastes.

Whole or crushed grains are the most common form of concentrated feed, sometimes referred to generically as "oats" or "corn" even if those grains are not present, also sometimes called "straights" in the UK.

Oats are the most popular grain for horses. Oats have a lower digestible energy value and higher fibre content than most other grains. They form a loose mass in the stomach that is well suited to the equine digestive system. They are also more palatable and digestible other grains.

Corn, referred to as Maize in the UK, is the second most palatable grain. It provides twice as much digestible energy as an equal volume of oats and is low in fibre. Because of these characteristics, is easy to over-feed corn, causing obesity, so horses are seldom fed corn all by itself. Nutritionists caution horse owners that moldy corn should never be fed because it is poisonous to horses.

Barley is also fed to horses, but needs to be processed to crack the seed hull and allow easier digestibility. It is frequently fed in combination with oats and corn, a mix informally referred to by the acronym "COB" (for Corn, Oats and Barley).

Wheat is generally not used as a concentrate. However, wheat bran is sometimes added to the diet of a horse for supplemental nutrition, usually moistened and in the form of a bran mash. Wheat bran is high in phosphorus, so must be fed carefully so that it does not cause an imbalance in the Ca:P ratio of a ration. Once touted for a laxative effect, this use of bran is now considered unnecessary, as horses, unlike humans, obtain sufficient fibre in their diets from other sources.

Many feed manufacturers combine various grains and add additional vitamin and mineral supplements to create a complete premixed feed that is easy for owners to feed and of predictable nutritional quality. Some of these prepared feeds are manufactured in pelleted form, others retain the grains in their original form.

In many cases molasses as a binder to keep down dust and for increased palatability. Grain mixes with added molasses are usually called "Sweet feed" in the USA and "Coarse mix" in the United Kingdom. Pelleted or *extruded* feeds (sometimes referred to as "nuts" in the UK) may be easier to chew and result in less wasted feed. Horses generally eat pellets as easily as grain. However, pellets are also more expensive, and even "complete" rations do not eliminate the necessity for forage.

The average modern horse on good hay or pasture with light work usually does not need supplements; however, horses subjected to stress due to age, intensive athletic work, or reproduction may need additional nutrition. Extra fat and protein are sometimes added to the horse's diet, along with vitamin and mineral supplements.

Soybean meal is a common protein supplement, and averages about 44 per cent crude protein. The protein in soybean meal is high-quality, with the proper ratio of dietary essential amino acids for equids. Cottonseed meal, Linseed meal, and peanut meal are also used, but are not as common.

Vegetable oil is a common fat source added to a ration. Corn oil is particularly popular, but other oils are used as well. Rice bran is a very good fat supplement that contains 20 per cent fat as well as fibre and other nutrients. Flax seed is another good source of fat, though it must ground up for horses to digest it. Some commercial feed manufacturers now make products containing both flaxseed and rice bran.

There are hundreds, if not thousands of commercially prepared vitamin and mineral supplements on the market, many tailored to horses with specialized needs.

Most horses only need quality forage, water and a salt or mineral block. Grain or other concentrates are often not necessary. But, when grain or other concentrates are fed, quantities must be carefully monitored. To do so, horse feed is measured by weight, not volume. For example, 1 lb. of oats has a different volume than 1 lb. of corn. When continuous access to feed is not possible, it is more consistent with natural feeding behaviour to provide three small feedings per day instead of one or two large ones.

However, even two daily feedings is preferable to only one. To gauge the amount to feed, a weight tape, available at most feed stores, can be used to provide a reasonably accurate estimate of a horse's weight. The tape measures the circumference of the horse's barrel, just behind the withers and elbows, and the tape is calibrated to convert inches or centimeters into approximate pounds or kilograms.

Actual amounts fed vary by the size of the horse, the age of the horse, the climate, and the work to which the animal is put. In addition, genetic factors play a role. Some animals are naturally easy keepers, which means that they can thrive on relatively small amounts of food and are prone to obesity and other health problems if overfed. Others are hard keepers, meaning that they are prone to be thin and require considerably more food to maintain a healthy weight.

Veterinarians are usually a good source for recommendations on appropriate types and amounts of feed for a specific horse. There are also numerous books written on the topic. Feed manufacturers usually offer very specific guidelines for how to select and properly feed products from their company, and in the United States, the local office of the Cooperative Extension Service can provide educational materials and expert recommendations.

Equids always require forage. When possible, nutritionists recommend it be available at all times, at least when doing so does not overfeed the animal and lead to obesity. It is safe to feed a ration that is 100 per cent forage (along with water and supplemental salt), and any feed ration should be at least 50 per cent forage. Hay with alfalfa or other legumes has more concentrated nutrition and so is fed in smaller amounts than grass hay, though many hays have a mixture of both types of plant.

When beet pulp is fed, a ration of 2 to 5 pounds is usually soaked in water for 3 to 4 hours prior to feeding in order to make it more palatable, and to minimize the risk of choke and other problems. It is usually soaked in a proportion of one part beet pulp to two parts water. Beet pulp is usually fed in addition to hay, but occasionally is a replacement for hay when fed to very old horses who can no longer chew properly.

Some pelleted rations are designed to be a "complete" feed that contains both hay and grain, meeting all the horse's nutritional needs.

However, even these rations should have some hay or pasture provided, a minimum of a half-pound of forage for every 100 pounds of horse, in order to keep the digestive system functioning properly and to meet the horse's urge to graze.

Recent studies address the level of various non-structural carbohydrates (NSC), such as fructan, in forages. Too high an NSC level causes difficulties for animals prone to laminitis or equine polysaccharide storage myopathy (EPSM). NSC cannot be determined by looking at forage, but hay and pasture grasses can be tested for NSC levels.

Concentrates, when fed, are recommended to be provided in quantities no greater than 1 per cent of a horse's body weight per day, and preferably in two or more feedings. If a ration needs to contain a higher percent of concentrates, such as that of a race horse, bulky grains such as oats should be used as much as possible; a loose mass of feed helps prevent impaction colic. Peptic ulcers are linked to a too-high concentration of grain in the diet, particularly noticed modern racehorses, where some studies show such ulcers affecting up to 90 per cent of all race horses.

In general, the portion of the ration that should be grain or other concentrated feed is 0 to 10 per cent grain for mature idle horses; between 20 to 70 per cent for horses at work, depending on age, intensity of activity and energy requirements.

Concentrates should not be fed to horses within one hour before or after a heavy workout. Concentrates also need to be adjusted to level of performance. Not only can excess grain and inadequate exercise lead to behaviour problems, it may also trigger Equine Exertional Rhabdomyolysis, or "tying up," in horses prone to the condition.

Horses normally require free access to all the fresh, clean water they want, and to avoid dehydration, should not be kept from water longer than four hours at any one time. However, water may need to be temporarily limited in quantity when a horse is very hot after a heavy workout. As long as a hot horse continues to work, it can drink its fill at periodic intervals, provided that common sense is used and that an overheated horse is not forced to drink from extremely cold water sources.

But when the workout is over, a horse needs to be cooled out and walked for 30—90 minutes before it can be allowed all the water it wants at one time. However, dehydration is also a concern, so some water needs to be offered during the cooling off process. A hot horse will properly rehydrate while cooling off if offered a few swallows of water every three to five minutes while being walked. Sometimes the thirst mechanism does not immediately kick in following a heavy workout, which is another reason to offer periodic refills of water throughout the cooling down period.

Even a slightly dehydrated horse is at higher risk of developing impaction colic. Additionally, dehydration can lead to weight loss because the horse cannot produce adequate amounts of saliva, thus decreasing the amount of feed and dry forage consumed. Thus, it is especially important for horse owners to encourage their horses to drink when there is a risk of dehydration; when horses are losing a great deal of water in hot weather due to strenuous work, or in cold weather due to horses' natural tendency to drink less when in a cold environment. To encourage drinking, owners may add electrolytes to the feed, additives to make the water especially palatable (such as apple juice), or, when it is cold, to warm the water so that it is not at a near-freezing temperature.

Special Feeding Issues for Ponies

Ponies and miniature horses are usually easy keepers and need less feed than full-sized horses. This is not only because they are smaller, but also, because they evolved under harsher living conditions than horses, they use feed more efficiently. Ponies easily become obese from overfeeding and therefore are at high risk for colic, Equine Metabolic Syndrome, and, especially, laminitis. Fresh grass is a particular danger to ponies; they can develop laminitis in as little as one hour of grazing on lush pasture.

It is important to track the weight of a pony carefully, by use of a weight tape. Forages may be fed based on weight, at a rate of about 1 pound of forage for every 100 pounds. Forage, along with water and a salt and mineral block, is all most ponies require.

If a hard-working pony needs concentrates, a ratio of no more than 30 per cent concentrates to 70 per cent forage is recommended. Concentrates designed for horses, with added vitamins and minerals, will often provide insufficient nutrients at the small serving sizes needed for ponies. Therefore, if a pony requires concentrates, feed and supplements designed specially for ponies should be used. In the UK, extruded pellets designed for ponies are sometimes called "pony nuts.".

Special Feeding Issues for Mules and Donkeys

Like ponies, mules and donkeys are also very hardy and generally need less concentrated feed than horses. Mules need less protein than horses and do best on grass hay with a vitamin and mineral supplement. If mules are fed concentrates, they only need about half of what a horse requires. Like horses, mules require fresh, clean water, but are less likely to over-drink when hot.

Donkeys, like mules, need less protein and more fibre than horses. They do best when allowed to consume small amounts of food over long periods, as is natural for them in an arid climate. They can meet their nutritional needs on 6 to 7 hours of grazing per day on average dryland pasture that is not stressed by drought.

If they are worked long hours or do not have access to pasture, they require hay or a similar dried forage, with no more than a 1:4 ratio of legumes to grass. They also require salt and mineral supplements, and access to clean, fresh water.

Like ponies and mules, in a lush climate, donkeys are prone to obesity and are at risk of laminitis.

Many people like to feed horses special treats such as carrots, sugar cubes, peppermint candies or specially manufactured horse "cookies." Horses do not need treats, and due to the risk of colic or choke, many horse owners do not allow their horses to be given treats. There are also behavioural issues that some horses may develop if given too many treats, particularly a tendency to bite if hand-fed, and for this reason many horse trainers and riding instructors discourage the practice.

However, if treats are allowed, carrots and compressed hay pellets are common, nutritious, and generally not harmful. Apples are also acceptable, though it is best if they are first cut into slices. Horse "cookies" are often specially manufactured out of ordinary grains and some added molasses. They generally will not cause nutritional problems when fed in small quantities. However, many types of human foods are potentially dangerous to a horse and should not be fed. This includes bread products, meat products, candy, and carbonated or alcoholic beverages.

It was once a common practice to give horses a weekly *bran mash* of wheat bran mixed with warm water and other ingredients. It is still done regularly in some places. While a warm, soft meal is a treat many horses enjoy, and was once considered helpful for its laxative effect, it is not nutritionally necessary.

An old horse with poor teeth may benefit from food softened in water, a mash may help provide extra hydration, and a warm meal may be comforting in cold weather, but horses have far more fibre in their regular diet than do humans, and so any assistance from bran is unnecessary. There is also a risk that too much wheat bran may provide excessive phosphorus, unbalancing the diet, and a feed of unusual contents fed only once a week could trigger a bout of colic.

All hay and concentrated feeds must kept dry and free of mold, rodent feces and other types of contamination that may cause illness in horses. Feed kept outside or otherwise exposed to moisture can develop mold quite quickly. Due to fire hazards, hay is often stored under an open shed or under a tarp, rather than inside a horse barn itself, but should be kept under some kind of cover.

Concentrates take up less storage space, are less of a fire hazard, and are usually kept in a barn or enclosed shed. A secure door or latched gate between the animals and any feed storage area is critical. Horses accidentally getting into stored feed and eating too much at one time is a common but preventable way that horses develop colic or laminitis.

It is also important to never give a horse feed that was contaminated by the remains of a dead animal, it is a potential source of botulism. This is not an uncommon situation. For example, mice and birds can get into poorly stored grain and be trapped; hay bales sometimes accidentally contain snakes, mice, or other small animals that were caught in the baling machinery during the harvesting process.

Horses can become anxious or stressed if there are long periods of time between meals. They also do best when they are fed on a regular schedule, they are creatures of habit and easily upset by changes in routine. When horses are in a herd, their behaviour is hierarchical; the higher-ranked animals in the herd eat and drink first. Low-status animals, who eat last, may not get enough food, and if there is little available feed, higher-ranking horses may keep lower-ranking ones from eating at all.

Therefore, unless a herd is on pasture that meets the nutritional needs of all individuals, it is important to either feed horses separately, or spread feed out in separate areas to be sure all animals get roughly equal amounts of food to eat. In some situations where horses are kept together, they may still be placed into separate herds, depending on nutritional needs; overweight horses are kept separate from thin horses so that rations may be adjusted accordingly.

Horses' teeth continually erupt throughout their life, are worn down as they eat, and can develop uneven wear patterns that can interfere with chewing. For this reason, horses need a dental examination at least once a year, and particular care must be paid to the dental needs of older horses. The process of grinding off uneven wear patterns on a horse's teeth is called *floating* and can be performed by a veterinarian or a specialist in equine dentistry.Diet and Nutrition for Dog

There are a great quantity of commercial foods and treats marketed for dogs, and not all are recommended as part of a balanced, healthy diet.

There is some debate as to whether domestic dogs should be classified as omnivores or carnivores, by diet. The classification in the Order Carnivora does not necessarily mean that a dog's diet must be restricted to meat; unlike an obligate carnivore, such as the cat family with its shorter small intestine, a dog is neither dependent on meat-specific protein nor a very high level of protein in order to fulfill its basic dietary requirements.

Dogs are able to healthily digest a variety of foods including vegetables and grains, and in fact dogs can consume a large proportion of these in their diet. Wild canines not only eat available plants to obtain essential amino acids, but also obtain nutrients from vegetable matter from the stomach and intestinal contents of their herbivorous prey, which they usually consume.

Domestic dogs can survive healthily on a reasonable and carefully designed vegetarian diet, particularly if eggs and milk products are included. Some sources suggest that a dog fed on a strict vegetarian diet without L-carnitine may develop dilated cardiomyopathy, however, L-carnitine is found in many nuts, seeds, beans, vegetables, fruits and whole grains. In the wild, dogs can survive on a vegetarian diet when animal prey is not available.

Observation of extremely stressful conditions such as the Iditarod Trail Sled Dog Race, and scientific studies of similar conditions has shown that high-protein (approximately 40 per cent) diets including meat help prevent damage to muscle tissue in dogs and some other mammals. This level of protein corresponds to the percentage of protein found in the wild dog's diet when prey is abundant; higher levels of protein seem to confer no added benefit.

Dogs frequently eat grass, which is a harmless activity. Explanations abound, but rationales such as that it neutralizes acid, or that dogs eat grass to induce vomiting to remove unwanted substances from their stomachs, are at best educated guesses. Dogs do vomit more readily than humans, as part of their typical feeding behaviour of gulping down food then regurgitating indigestible material such as bones and fur.

This behaviour is typical of pack feeding in the wild, where the most important thing is to get as much of the kill as possible before others consume it all. Individual domestic dogs, however, may be very "picky" eaters, in the absence of this social pressure.

Dogs may also appear to eat grass when they are just running the blades through their mouth to gather information. Their sense of smell and taste may act together to detect if other animals have walked through their area or urinated on the grass.

DANGEROUS SUBSTANCES

Human food

Some foods commonly enjoyed by humans are dangerous to dogs, including chocolate (Theobromine poisoning), onions, grapes and raisins, some types of gum, certain sweeteners and Macadamia nuts. The only known dangerous substance in chocolate is cocoa, so the danger of white chocolate is uncertain. The acute danger from grapes and raisins was discovered around 2000, and has slowly been publicized since then. The cause is not known. Small quantities will induce acute renal failure.

Sultanas and currants may also be dangerous. Cooked bones are dangerous for dogs, because the heat of cooking changes their chemical and physical properties so that they cannot be chewed properly. As a result they may splinter into jagged shards that resist digestion. Alcoholic beverages pose comparable hazards to dogs as they do to humans, but due to low body weight and lack of alcohol tolerance they are toxic in much smaller portions.

Plants

Plants such as caladium, dieffenbachia and philodendron will cause throat irritations that will burn the throat going down as well as coming up. Hops are particularly dangerous and even small quantities can lead to malignant hyperthermia. Amaryllis, daffodil, english ivy, iris, and tulip (especially the bulbs) cause gastric irritation and sometimes central nervous system excitement followed by coma, and, in severe cases, even death.

Ingesting foxglove, lily of the valley, larkspur and oleander can be life threatening because the cardiovascular system is affected. Yew is very dangerous because it affects the nervous system. Immediate veterinary treatment is required for dogs that ingest these.

Household Poisons

Many household cleaners such as ammonia, bleach, disinfectants, drain cleaner, soaps, detergents, and other cleaners, mothballs and matches are dangerous to dogs, as are cosmetics such as deodorants, hair colouring, nail polish and remover, home permanent lotion, and suntan lotion. Dogs find some poisons attractive, such as antifreeze (automotive coolant), slug and snail bait, insect bait, and rodent poisons. Antifreeze is insidious to dogs, either puddled or even partly cleaned residue, because of its sweet taste. A dog may pick up antifreeze on its fur and then lick it off.

Animal Feces

Dogs occasionally eat their own feces, or the feces of other dogs and other species if available, such as cats, deer, cows, or horses. This is known as coprophagia. Some dogs develop preferences for one type over another. There is no definitive reason known, although boredom, hunger, and nutritional needs have been suggested. Eating cat feces is common, possibly because of the high protein content of cat food. Dogs eating cat feces from a litter box may lead to to Toxoplasmosis. Dogs seem to have different preferences in relation to eating feces. Some are attracted to the stools of deer, cows, or horses.

Other risks

Human medications may be toxic to dogs, for example paracetamol/acetaminophen (Tylenol). Zinc toxicity, mostly in the form of the ingestion of US pennies minted after 1982, is commonly fatal in dogs where it causes a severe hemolytic anemia.

COW DIET

Corn has been a staple ingredient in lactating dairy cow rations for a hundred years. However, many dairy farms import corn to feed to their dairy cows. The price of this imported corn has been fairly stable for a number of years.

The government is trying to reduce dependence on foreign oil by supporting efforts to produce fuel (ethanol) from renewable resources like corn or cellulose. There is widespread public support for incorporating ethanol into fuelburning vehicles (gasoline and diesel) because ethanol addition results in greater combustion of these fuels and is nontoxic and biodegradable in the water and the soil.

The Congress passed legislation called a renewable fuels standard (RFS) that will at least double the use of ethanol and biodiesel by the year 2012. The number of plants producing ethanol from corn is projected to increase dramatically.

This growth in ethanol production has reduced the amount of corn available for livestock feed and concurrently increased its market price. Increased corn prices have resulted in increased feed costs. While higher milk prices may compensate for ration cost increases, partially replacing corn with other less expensive feedstuffs may be required over the longer term.

ALTERNATE RATION STRATEGIES

Other Grain Sources of Dietary Starch

Grains, like corn, are an excellent source of starch which is highly fermentable by ruminal microorganisms. The propionic acid produced by starch-fermenting bacteria is converted to glucose by the cow's liver; this glucose is used to make milk. In addition, the bacteria provide about 50 to 60 per cent of the protein needs of the cow as they are washed out of the rumen and are digested in the abomasum and small intestine.

Fig. Grain Diet for Cow

To optimize milk production, starch should make up 24 to 26 per cent of the dietary dry matter. Barley and wheat are rich sources of dietary starch. Shifting cow diets between corn and barley or wheat needs to be done slowly.

The starch in barley and wheat is more rapidly available than that in corn so quick dietary changes can result in digestive upsets and lowered ruminal pH. Wheat flour can sometimes be bought for less than corn, but it is very dusty and potentially explosive so it is often avoided by mills. Field reports suggest that flour might 'paste up" in the rumen, so it is not a popular feed ingredient. Hominy contains less starch but more protein, fibre, and fat compared to corn, however their energy density is similar.

Replacing all of the corn with hominy will reduce the dietary starch concentration about 3 percentage units. Dietary costs would be expected to be reduced by replacing corn with hominy. If no milk is lost, then profit is improved. Experiments comparing performance of cows fed ground corn versus hominy could not be located. However, many Florida dairies feed hominy successfully. Even replacing half the corn with hominy can be a reasonable strategy. Because hominy is higher in fat and phosphorus than corn, care should be taken to avoid overfeeding these nutrients.

Low Starch Feedstuffs as Possible Substitutes for Corn

Because grain sources rich in starch, such as barley or wheat, may not be economically available year-round, other commodity feeds must be considered.

Ingredients with lower prices than corn should be considered as partial replacements for corn. Most common feedstuffs contain quite a bit less starch than corn. The best starch sources after corn include corn silage at 30 per cent, wheat midds at 26 per cent, wheat bran at 23 per cent, rice bran at 19 per cent, and corn gluten feed at 16 per cent (DM basis). All other feedstuffs are less than 10 per cent starch. Replacing some of the corn with these feeds will reduce dietary starch to less than 25 per cent which is a common, US ration target. How far can dietary starch be reduced without significantly affecting milk production?

Wheat Midds

Wheat midds contain 26 per cent starch and 18 per cent protein. Price can be substantially lower than corn at the right time of the year. Wheat midds have not been evaluated as a substitute for corn in many experiments. In a recent study, wheat midds were fed at about 7.5 per cent of the diet, replacing a combination of corn and soybean meal.

Corn was reduced in the diet from 34.1 to 28.9 per cent. Cows fed wheat midds tended to eat less feed dry matter·but milk production and milk composition were not affected. This reduced feed intake may have been because wheat midds have a high water-holding capacity, thus increasing gut fill with fibre. Cows fed wheat midds appeared to have looser manure than those fed more corn and soybean meal.

In a second study, wheat midds were fed at 0 or 22.4 per cent of the diet, replacing 35 per cent of the ground corn and 30 per cent of the soybean meal. Intake of feed dry matter, production of milk, and milk composition were not different between the two groups of cows which averaged 150 days in milk at the start of the study. However the digestibility of dry matter and neutral detergent fibre were lower for cows fed wheat midds. Therefore wheat midds do not appear to be an effective feed replacement for ground corn except for lower producing cows.

Corn Feed

Corn gluten feed (CGF) is a byproduct of the manufacture of corn sweeteners, corn starch, corn syrup, and corn oil using the wet milling process. The corn starch is used to make ethanol.

The CGF consists of the corn bran and a steep liquor (fermented nutrients extracted from water used to soak corn grain) mixed in approximately a 2:1 ratio. CGF is sold in both wet and dry forms. It contains 16 per cent starch, 36.1 per cent NDF, and 23.5 per cent protein, the protein being highly degradable in the rumen.

Pricing is substantially lower than corn. CGF may serve as a concentrate, replacing only the corn and soybean meal, or as both a concentrate and a fibre source, replacing both concentrate and traditional forages. CGF generally accounted for 10 to 45 per cent of the diet, although some studies went higher.

When CGF was fed at about 20 per cent of ration DM, milk production was decreased statistically in 1 study, increased in 1 study, and remained unchanged in 9 other studies. Cows in these studies were milking between 50 and 90 pounds per day.

Increasing the CGF to 30, 40 or even 57 per cent of the diet did not reduce milk in any study with the exception of Staples et al. In two studies, in which CGF was fed between 38 and 40 per cent of the diet replacing both forage and concentrate starting at calving, average milk yield was increased significantly by feeding the corn gluten feed.

As more CGF replaced more corn, the dietary starch concentrations dropped, going as low as 15 per cent in some cases. In spite of lowered starch intake, milk yield did not drop significantly.

The bulk of these studies indicate that dietary starch could be reduced by replacing some corn and protein meals with the digestible fibre found in CGF.

Dropping dietary starch from 26 to 21 per cent may be an acceptable compromise. Sample diets were formulated to include CGF at 0 per cent, 10 per cent, and 20 per cent of the diet; starch was reduced concurrently from 25.9 to 23.5 per cent to 20.9 per cent in these rations. As CGF increases in diet, the proportion of extruded soy meal increased to keep the ruminally undegradable protein constant.

Degradable protein sources had to be reduced to keep the dietary crude protein from exceeding 17 per cent. Whether this approach will maintain milk production is not known. Milk and feed intake will be maintained when CGF is fed at 20 per cent of dietary DM. But feeding CGF at 10 per cent of the diet may be a preferred approach initially to "play it safe."

Some signs that may indicate that the dietary starch has gotten too low include; dropped milk production, stiffer manure, increases in milk urea nitrogen, and loss of body condition.

Another concern is that drying CGF can reduce the digestibility of the protein if the drying temperature is too high. This is determined by analyzing CGF for acid detergent insoluble protein (ADIN). Bernard et al.reported that dried CGF had higher concentrations of ADIN than wet CGF. In recent years, high ADIN may not be as big of a problem in CGF based upon analyses reported by Dairy One. The ADIN values were less than 5 per cent of the nitrogen in the majority of samples; ADIN is usually not considered a problem until it reaches 10 per cent of the total nitrogen in the feed.

Commodities need to have some consistent level of nutrient density from delivery to delivery in order to keep the daily ration nutrients consistent for the cows. Dairy One lists on their web site the normal range and standard deviation of the nutrients of most of the feeds they have analysed over the past 6 years.

Corn has an average starch value of 70.6 per cent. The standard deviation is 5.1 per cent. This means that two-thirds of the samples that were analysed ranged between 65.5 per cent and 75.7 per cent which is 5.1 per cent units added to or subtracted from the mean of 70.6 per cent.

This also means that one-third of the samples analysed were outside this range. This is a fairly narrow range compared to byproduct feeds. Starch in CGF samples has a typical range of 8.6 per cent to 24.0 per cent with a mean of 16.3 per cent.

This large variation is likely due to the fact that CGF production is not standardized across mills. Likewise, starch in hominy typically ranges between 42.9 per cent and 63.9 per cent with a mean of 53.4 per cent. By feeding these byproducts in smaller amounts, the potential negative effect of nutrient variation on cow performance is reduced. When contracting for these commodities, an acceptable range in variation between loads should be agreed upon ahead of time. Loads delivered outside this agreed-upon-range would then be refused.

Dried Distillers Grains Plus Solubles (DDGS)

Dry corn put through the dry-milling process will produce ethanol, carbon dioxide, and DDGS. Each 56-pound bushel of corn processed through the dry milling process results in 2.85 gallons of ethanol, 18 pounds of carbon dioxide, and 18 pounds of DDGS. Plants using the dry milling process are less expensive to build than the plants using the wet milling process, thus about 75 per cent of the ethanol produced from corn comes from dry milling.

The production of DDGS in the US increased tenfold between the years 1980 and 2000, increasing from 320,000 to 3.5 million metric tons (1 metric ton = 2205 pounds). Production doubled again between 2000 and 2004 to over 7.3 million metric tons. Every dairy cow in the U.S. would need to consume 7.3 pounds per day of DDGS over a 305-day lactation in order to use up this supply, but of course DDGS is also fed to other livestock and is exported.

During the production of ethanol from corn, a syrup product and a cake product are produced. These products are blended together in different proportions to form DDGS. Depending on the plant, the DDGS may be a mixture of syrup to cake ranging in proportion from 35:65 to 55:45. Because of the variation in the ratio of syrup to cake used by different plants to form DDGS, nutrient concentrations will vary from one plant to another.

However the variation looks to be less than that of CGF. Although DDGS is very low in starch, it is an excellent source of protein (30.3 per cent) and fat (13 per cent) and therefore has an energy density very similar to corn. In addition, the ethanol-making process increases the digestibility of the fibre. Unlike CGF, DDGS is a very good source of ruminally undegradable protein (RUP) but if the temperature of the drying process is too high, the protein may be indigestible.

Purchasers of DDGS should analyse for ADIN regularly. Kalscheur et al. reviewed 24 experiments in which 98 comparisons were made between cows fed diets containing DDGS and those not fed DDGS. Cows fed DDGS at up to 30 per cent of the ration produced as much milk as those not fed any DDGS (about 73 lb/day). In spite of decreasing starch content, milk production was maintained. In a study conducted at the University of Florida, whiskey DDGS were fed at 0 per cent or 20 per cent of a 55 per cent concentrate:45 per cent forage diet.

The forage was all corn silage in one set of diets and 50 per cent corn silage:50 per cent rhizome peanut silage in another set of diets. The DDGS replaced around 40 per cent of the corn and soybean meal. Milk production was increased from 59.0 to 60.7 pounds per day without changing milk fat test.

The effect of DDGS was the same regardless of whether the forage was totally corn silage or an equal mixture of corn silage and rhizome peanut silage. However, a maximum feeding level of DDGS might be 15 per cent in order to prevent the overfeeding of protein, RUP, unsaturated fat, and phosphorus. Increased corn oil in the DDGS can also lead to decreasing milk fat, so fat content of DDGS should be monitored regularly.

Feeding DDGS at 20 per cent of the diet will increase dietary fat by 2.6 per cent. This will likely be a concern if the diet contains other supplemental fats. Dietary fat should be kept below 6 per cent. The feeding of one pound of DDGS can replace about 0.6 pounds of corn and 0.4 pounds of soybean meal.

Soybean Hulls (SBH)

During the processing of soybeans for oil and meal, the hulls are separated and ground. The resulting soybean hulls consist largely of the outer covering of the soybean so they are high in fibre and contain moderate amounts of protein but very little starch. Protein concentration will depend upon how well the hulls are cleaned. Positive characteristics include a high content of lysine, a low content of phosphorus, and a highly digestible fibre. Fat content can vary widely.

The availability of SBH is likely to increase in coming years as more acres are planted to soybeans in order to use soybean oil for biodiesel production. In 13 separate experiments published between 1976 and 2002, SBH (fed at up to 20 per cent of the dietary DM) successfully replaced ground corn or high moisture corn in rations for lactating cows. In most of these studies, the dietary starch levels were probably greater than what is recommended today, based upon the proportion of corn in the diets fed in these studies. Therefore the replacement of corn with SBH still left enough starch to support adequate intake and milk production.

One study fed a more conservative amount of corn in the control diet and deserves further attention here. The control diet was 23 per cent high moisture corn, 26 per cent corn silage, and 26 per cent alfalfa silage. Adding SBH to the diet at 14 per cent of the dietary DM reduced the corn to 9 per cent of the diet.

This dropped the starch from roughly about 25 per cent to 15 per cent, calculated from average values. Diets were fed starting at 8 days fresh.

Lactating cows ate more DM when offered the ration containing SBH but milk production was not changed. Lactating heifers fed SBH ate the same amount of feed and produced the same amount of milk as those not fed SBH. Concentrations of milk fat and protein were unchanged. Replacing 60 per cent of the corn with SBH, so that SBH made up 14 per cent of the ration, was effective to support high milk production in early lactation.

Even though dietary NDF increases when SBH replace corn, feed intake has not been decreased. This is in contrast with the well-documented fact that as traditional forage replaces concentrate in the diet, concentration of dietary NDF increases and feed intake decreases. However the NDF in nonforage fibre byproducts like SBH does not have the same physical characteristics as the NDF in traditional, properly chopped forage.

The fibre in SBH is highly digestible, very short and has a specific gravity that allows it to move out of the rumen quickly so that SBH fibre does not have the same rumen fill as forage fibre at similar NDF concentrations. In order to use SBH most efficiently, the diet should contain sufficient "effective" fibre from traditional forages.

Corn Silage

Corn silage hybrids differ in the proportion of grain in the total crop. Selecting corn silage hybrids that contain more starch will reduce the amount of ground corn needed. In a test of 55 corn silage hybrids grown in Gainesville in 2006, starch concentration ranged from 22 per cent to 35 per cent of silage dry matter, with an average of 29.3 per cent.

Selecting a corn silage hybrid with high levels of starch and digestible fibre, while maintaining high, could be a good strategy to reduce corn costs in the ration. How much could be saved?

If a corn silage hybrid had 33 per cent starch instead of 30 per cent, and corn silage was fed at 33 per cent of the ration dry matter, dietary ground corn could be reduced from 18 per cent to 16.5 per cent, which is about 1 lb/day less ground corn.

The cost savings would depend upon the relative price of the feedstuff(s) used.

Corn silage contains much more starch than other forages fed in Florida. Feeding more corn silage in the ration and less sorghum silage, cottonseed hulls, or bermudagrass will increase the starch in the diet and allow for less ground corn to be fed. If corn silage inventory allows, increasing the corn silage from 33 to 38 per cent of the diet DM by replacing another forage like sorghum silage, will allow ground corn to be reduced by 1.5 percentage units.

Glycerol

Glycerol is a byproduct made as plant and animal fats are processed to make diesel fuels. Diesel fuel produced from fat burns cleaner, as there is no soot or particulate matter produced. Starting with animal fat to make diesel fuel is more profitable than starting with vegetable oil because animal fats are cheaper.

However, diesel made from animal fat may not work as well in colder climates because it "clouds up." This process is expanding worldwide and therefore the supply of glycerol in the future is likely to increase. Ten pounds of glycerol result from every 100 pounds of biodiesel produced.

The purity of this byproduct can vary widely. The more impure the product, the more water, methanol, phosphorus, and potassium it contains. A product with these contaminants is labeled glycerin. The glycerin fed in a German study contained 2.2 per cent potassium and up to 2.4 per cent phosphorus. Methanol is present due to its use in the manufacture of biodiesel. Although ruminal bacteria can detoxify methanol, it can be problematic if present in large quantities.

According to the FDA, glycerin is considered a substance that is GRAS (Generally Recognized as Safe) for general purpose use in animal feed, unless methanol is present at concentrations exceeding 150 ppm. The energy content of glycerol is similar to corn when fed in high starch diets. Dietary glycerol would be converted mainly to propionate and butyrate by ruminal bacteria. As a feed ingredient, it could substitute for corn or molasses.

Little research has been done regarding the feeding of glycerol to dairy cows. Lactating dairy cows fed glycerol (1.3 per cent methanol) at either 0 per cent, 2.7 per cent, or 5.3 per cent of dietary dry matter ate the same amount of feed and produced the same amount of milk. Milk composition was unchanged with the exception that milk urea nitrogen concentrations were lower in cows fed glycerol.

Efficiency of fat-corrected milk production was improved from 1.46 to 1.59 and 1.60 lb of FCM yield per lb of feed dry matter intake. Interestingly, animals of other species consumed more water when they were fed glycerol. This may have a benefit to dairy cows managed under heat stress conditions. German researchers fed glycerol up to 10 per cent of ration DM successfully. Glycerol appears to be a good pelleting agent when added at 5 per cent.

Future Byproducts from new Technologies to Improve Ethanol Production

As the ethanol manufacturing industry works at improving the efficiency of conversion of corn to ethanol, different byproducts will become available. Applegate et al. lists the following potential corn byproducts.

- Called the "quick germ quick fibre method," an enzyme is added to the water used to soak the ground corn, causing the germ and fibre to float before the fermentation process begins. The product from this process contains 28 per cent protein, 5 per cent fat, and 25 per cent NDF.
- Collecting the pericarp fibre and germ prior to fermentation by modifying the dry grinding process (drum degerminator) results in a product that is 24 per cent protein, 8-9 per cent fat, and 28 per cent NDF.
- Removing the fibre by sieving and air aspiration results in a product that is 40+ per cent protein, 15 per cent fat, and 20 per cent NDF.

- Modifying the yeasts used to ferment the sugar to ethanol could allow for an increased concentration of lysine in DDGS, thus giving DDGS a more favourable amino acid profile.

As each mill adopts what they consider the best ethanol-producing technology, the feeding industry will be faced with a variety of byproducts on the market which will need to be properly identified prior to purchase. With the increased price for corn due to demand for ethanol, farmers will be planting more of their acres to corn. This will reduce the number of acres committed to other crops, such as cottonseed.

This shift will likely have a large impact on the market price of a number of feed commodities. The availability and price of each commodity will need to be evaluated for optimal pricing. Some acceptable, alternative feeds may allow some reduction of corn grain in the diet of lactating dairy cows.

FISH FEEDING

Providing fish with the right type of food in the right proportions is a very important task. This can be much more of a challenge in a saltwater tank than in a freshwater tank, because most marine fish are still wild-caught, and their natural diet can be difficult to maintain in captivity. With the right information, however, most marine fish will learn to eagerly anticipate their feeding time and will voraciously devour their food.

A good rule of thumb to follow is that fish should be fed no more than they can consume within a five minute period. Any food that is left behind after feeding time should be removed before it has a chance to begin decomposing. If it is found that fish are consuming the food in far shorter a time than five minutes and seem anxious for more, the amount that are fed must be increased.

On the other hand, if fish are healthy, but they are only consuming about half of what they are fed, probably need to start feeding less. Always keep in mind that much more harm can be done to fish by overfeeding than by underfeeding. In the wild, reef fishes will browse for food throughout the day.

Their natural habitat as closely as possible, it is recommended that fish must be fed several small meals every day. At the very least, feed two medium-sized meals per day, one in the morning and one at night.

Types of Fish Food

The kind of food to feed depends on the kind of fish are feeding: herbivore, carnivore, or omnivore.

Herbivores

It is common for fish that graze on algae in the wild to suffer in captivity because they don't receive the appropriate type of food to keep them healthy. Many of these fish are fed terrestrial greens like spinach and romaine lettuce, but these are not appropriate foods for these fish that are solely algae eaters.

They should be fed either dried or fresh algae. Dried algae is sold in paper-thin sheets that can be broken down into appropriate portion sizes, depending on the number of fish are feeding. Fresh algae can be obtained by placing a glass container, with one rock in aquarium, on a sunny windowsill. Wait for the rock to grow a nice layer of algae, and then place it back in the aquarium for fish to graze on.

Herbivorous fish can also be fed one of the frozen fish foods that is formulated especially for herbivores. This food should contain a variety of marine algae, vegetable matter, and seafood. Whether herbivorous fish are fed dried algae, fresh algae, or a frozen food formula, herbivorous eaters are grazers and should have food available to them at all times, unlike most other fish.

Carnivores

Meat-eating marine fish can enjoy a variety of fresh, frozen, freeze-dried, and live foods. Fresh foods suitable for fish include many seafood items, such as shrimp, scallops, mussels, clams, and some fish. If seafood is fed to fish, cut it into bite-sized pieces (or serve it whole to larger fish). Fresh seafood can be frozen and then thawed as needed.

Frozen foods are available from most fish shops and other sources. These offerings may include brine shrimp, bloodworms, and daphnia. They may be in the form of a slab that can be broken into smaller pieces or small, ready-to-go cubes. To feed frozen foods to carnivorous fish, thaw a small portion in aquarium water, and then drop it in the tank.

Several freeze-dried foods for carnivorous fish are readily available, such as plankton, krill, brine shrimp, and bloodworms. Freeze-dried foods are foods that have been completely dehydrated, yet they maintain most of their nutritional value.

They have a relatively long shelf life, particularly when compared to fresh and frozen foods. Freeze-dried foods should be used as supplements for carnivorous fish. Live foods are almost always a favourite of fish. Adult brine shrimp, bloodworms, and Mysis shrimp are among the better choices. Earthworms are also a great live food choice because they come in all sizes, fish love them. Be sure to thoroughly rinse them before feeding them to fish.

Omnivores

The majority of marine fish are omnivorous, which means they need to eat both meat- and plant-based foods. One easy option for omnivorous eaters is commercial fish food, such as flakes or pellets. However, offering a varied diet will give healthier, more colourful fish. Try to feed fish two feedings a day of commercial fish food and one feeding of a meaty item, such as bloodworms or chopped fish. The food offered must be alternated in addition to commercial food to keep the fish from getting bored and to make sure all of their nutritional needs are met.

SHEEP DIET

Sheep are exclusively herbivorous mammals. Like all ruminants, sheep have a complex digestive system composed of four chambers that allows them to break down otherwise inedible cellulose from stems, leaves, and seed hulls in to complex carbohydrates.

When sheep graze, vegetation is chewed in to a mass called a bolus, which is then passed in to the first chamber: the rumen. The rumen is a 5-10 gallon (18.9-37.8 liter) organ which ferments feed via a symbiotic relationship with the millions of bacterium, protozoa, and yeasts which are collectively called flora.

The bolus is periodically regurgitated back in to the mouth as cud for additional chewing and salivation, this process usually takes around six hours. Cud chewing is an evolutionary adaptation that allows ruminants to graze more quickly in the morning, and then fully chew and digest feed later in the day. This is beneficial because grazing, which requires lowering the head, leaves sheep more vulnerable to predators, while cud chewing does not.

During fermentation, the rumen produces excess gas which must be expelled; disturbances of the organ, such as sudden changes in a sheep's diet, can cause potentially fatal conditions such as bloat. After fermentation in the rumen, feed passes in to the reticulum and the omasum; special feeds such as grains may bypass the rumen altogether.

Following the first three chambers, food moves in to the abomasum for final digestion before processing by the intestines. The abomasum is the only one of the three chambers directly equatable with the usual mammalian stomach, and is accordingly sometimes called the "true stomach".

Sheep follow a diurnal pattern of activity, beginning to feed at dawn and throughout the day until dusk, stopping sporadically to rest and chew their cud. Ideal pasture for sheep is not lawn-like grass, but an array of grasses, legumes and forbs. Types of land where sheep are raised vary widely, from pastures that are seeded and improved intentionally to rough, native lands.

Common plants toxic to sheep are present in most of the world, and include (but are not limited to) oak and acorns, tomato, yew, rhubarb, potato, and rhododendron. Sheep also may perish from consuming man-made substances.

Sheep are largely grazing herbivores, unlike browsing animals such as goats and deer that prefer taller foliage. With a much narrower face, they can crop plants very close to the ground and can overgraze a pasture much faster than cattle. For this reason, many shepherds utilize some type of managed grazing, whereby a flock is rotated through multiple pastures, giving plants time to recover.

Paradoxically, sheep can be both the cause of and solution to one of the major ill effects of overgrazing: the spread of invasive plant species. By disturbing the natural make-up of pasture, sheep and other livestock can pave the way for invasive plants. But sheep also prefer to eat invasives such as cheatgrass, leafy spurge, kudzu and spotted knapweed over native species such as sagebrush—thereby making grazing sheep an effective measure in restoring native pastures.

Other than forage, the only other common staple feed for sheep is hay, most often during the winter months. The ability to survive solely on pasture (even without hay) varies with breed, but all sheep are capable of doing so. Also included in most sheep's diets are minerals, either in a trace mix or in licks.

Naturally, a constant source of potable water is also a fundamental requirement for sheep. The amount of water needed by sheep fluctuates with the season and the type and quality of the food they consume. When sheep feed on large amounts of new growth and there is precipitation (including dew, as sheep are dawn feeders), sheep need less water. When sheep are confined or are eating large amounts of cured hay, more water is typically needed. Sheep also require clean water, and may refuse to drink water that is covered in scum or algae.

COOLING PONDS FOR DAIRY CATTLE

Cattle have used water for cooling probably for as long as there have been cattle. The last survey done in Florida reported that 30 per cent of the dairies had cooling ponds. The reader should be aware that cooling ponds, if not maintained properly, could harbor infectious diseases such as Leptospirosis and many mastitis organisms. The water usage is very high to fill and keep full.

Environmental concerns that must be addressed are the run-off from ponds and ground seepage. Every state seems to have many regulations that must be addressed before building cooling ponds. The benefits of cooling ponds and some suggestions on building and maintaining them. It is the dairymen's responsibility to use them properly and to be aware that ponds can harbor diseases and the pond water must be dealt with according to state and local laws.

What is a cooling pond for dairy cattle? For this discussion they are a man made hole in the ground with a constant inflow of clean fresh water. Usually the ponds drain or overflow from the top. If a series of ponds is used, the overflow water may be run into a retention pond to be irrigated on crops or enter the waste management system. Cooling ponds may also be fed from springs or artesian wells. So cooling ponds for dairy cattle should have constant fresh water in and a constant outflow of water to some place for proper disposal.

Cement ponds may also be built. These may drain from the bottom and can be used to flush feed lanes or barns. Bottom draining will help remove some of the solids that have settled to the bottom of the pond. Care must be taken on the entrance and exit to cement ponds.

These may be very slippery if the cement is not cross-grooved. A stagnant pond without fresh water entering, a lagoon, a wallow, and a mud hole do not classify as a cooling pond for dairy cattle.

Site Selection

A site near the feed and drinking water, ideally with some form of shade for cows, is best. The idea is for cows to cool, leave the pond and goes eat and drink, lay down in the shade and then do it all over again. If cows do not have feed and drinking water near, they will stay in the pond and not eat and we have cool wrinkled cows that do not give much milk.

Cooling ponds should be located close to a water supply as fresh water needs to enter the ponds constantly. There must be a way to dispose of runoff water properly. This will vary by areas of the country.

Some slope from the ponds to the waste management area would be helpful, so pumps need not to be used. Cooling ponds should not be too far from the milking parlor as long walks in the sun increase heat stress. Exit lane sprinklers help make the long walk more comfortable.

Concrete Ponds

Concrete ponds will work and have been used but they are expensive to build. However, they are easier to maintain, as the entrances and exits do not have to be rebuilt. The finish of the concrete on exit and entrance slopes should be cross-grooved to prevent cows from falling. A 1:8 slope would be a safe slope if concrete ponds are constructed.

Sizing Cooling Ponds

There is no official listing for cooling pond size. If a holding area requires $15/ft^2$ / cow and cows are driven into a pond and taken out as a group then $15/ft^2$ / cow might be adequate. If $50/ft^2$ / cow is recommended for most general animal space, then this might be the correct figure for a cooling pond.

This amount of space would allow cows to enter and exit the pond at all times when all cows were not using the pond at the same time. If $50/ft^2$ / cow for a pond is used, then 5000/ ft^2 is then needed for the 100 cow group.

Pond Shapes and Depths

Using the sample of $5000/ft^2$ /100 cows a pond could be 50' wide × 100' long, or a circle 80' in diameter (A= (Pi) r^2; A= 3.14 × 40 = $5024ft^2$). If a 50' × 100' pond is used and the sides are fenced so the cows can only enter or exit at each end of the pond, the amount of dirt falling into the pond will be reduced. If a circular pond is made, cow carpet should be used to eliminate the problem of dirt falling in the pond. As cows can swim, pond depth is not a problem.

Deeper ponds may allow the settling of organic matter on the bottom of the pond. This organic matter may not be disturbed by cows walking through it as they will swim over the top of it. Preference should be given to depths of 3' or 4' as it seems to cool the cows and let them stand and walk in and out at will. While deep ponds allow cows to submerge, they do not seem to like to float too long so they go to where they can stand and may block cow traffic. A circular pond that is deep in the middle lets cows dunk themselves while others enter and exit the pond at will.

Water Usage

A 100 cow group with $50/ft^2$ / cow could have a 50' wide × 100' long × 3' deep cooling pond (it may be 4' in the middle and slope in and out). A cooling pond 50' wide × 100' long × an average 3' deep=$15000/ft^3$. There are 7.48 $gallons/ft^3$, thus the pond requires 112,200 gallons of water.

The pond must be filled and then kept fresh with running water at all times while in use. A 1" pipe 300 ft. long and water pressure at 35psi will result in a flow rate of about 23.5 gpm. At this rate it would then take 80 hours to fill the pond initially.

The dairy water usage would then be 1,410 gph × 24 hours or 33,840 gallons of water per day. This fresh water must be pumped into the pond and there is going to be close to that amount of run off water that will exit the pond for disposal.

Water Run-off

Disposal of run-off water must be in accordance with whatever local, state, and federal laws that applies. Many dairies run the water into the waste management system, while others add it to retention ponds to be used for irrigation.

Shading over Ponds

Experience has led to discontinued use of shade over ponds as the cows never leave the ponds. Remember, the object of this exercise is to increase dry matter intake and cows will not eat much soaking in the pond.

Pond Maintenance

At least once a year, usually during the winter, the pond should be fenced and the water removed by pumping. The mud and manure on the bottom of the pond should be removed and the pond left empty for the sun to shine on it.

In the spring, the pond should be filled again and the fences removed. Some people just fence the pond in the winter, dig a new pond and fill in the old one with new pond's dirt.

Mycoplasma mastitis is often cultured in water from cooling ponds. This is a real problem if pulsators do not work well resulting in damaged teat ends. Many herds with mycoplasma positive ponds, have no mycoplasma in the bulk tank.

If the water is filthy, cows should not be permitted in the ponds and the ponds should be pumped dry. The sludge should be cleaned and the ponds filled with water again. Both the sludge and the water shoud be removed from cow areas.

Cooling Ponds to Reduce Body Temperature

In the summer of 1986, an experiment was conducted at a North Florida dairy that was using man-made cooling ponds. Ten early lactation Holstein cows were fitted with radio transmitters that transmitted inner ear temperatures every five minutes to a radio receiver and data logger.

Data were collected over a four-day period in mid-August. The cows' activities were also monitored during this time. Entering and exiting the ponds, eating, drinking and laying were recorded.

The cooling ponds lowered the cows' temperature by 1-2°F depending on the time of day they entered the cooling ponds. The average length per stay in the pond was 18 minutes for events from midnight to noon and 12 minutes per visit from noon to midnight. It was obvious that cooling ponds cooled cows.

Pond Water Quality

In 1987, samples were taken from a dairy's man-made earthen ponds. The ponds were sampled weekly from May 19,1987 to September 28, 1987. Six to ten ponds were sampled each week.

Total bacteria counts (TBC), varied greatly from week to week. This may have been due to no water entering the ponds. There did not seem to be any great increase in total bacteria counts or coliform counts as time progressed. The total bacteria count averaged 3,133,700 CFU/ml for all ponds for the 20 weeks. The coliform counts averaged 14,340 CFU/ml for the same period.

Cooling Ponds and Their Effect on Milk Quality

In a previous study on a dairy in Florida with man-made cooling ponds, it was reported that cows exposed to cooling ponds did not experience more clinical mastitis than cows that did not have access to cooling ponds. In fact, cows exposed to cooling ponds during the trial period were only half as likely to develop a case of clinical mastitis as cows not exposed to cooling ponds.

Clinical Mastitis Organisms

The same dairy had provided its clinical mastitis records for several years. The total number of clinical mastitis cases was high in the first quarter of the year and then declined sharply about the time hot weather came and the cows started using the ponds. The incidence stayed low, even in the last quarter of the year when the ponds were not in use. This can be explained many ways. The first quarter of the year was an extremely wet period.

The cows were quite dirty and the cow wash system was available to one-half of the herd. This may explain some of the variation, even though one-half of the herd had never had a cow wash. Another confounding variable was that this herd switched to Clorox for pre- post-teat dipping in April of that year. It had previously pre- and post-dipped with a Chlorhexidine product.

From this it could be concluded that Clorox caused the reduction of clinical mastitis. However, there were no controls in this study, one could conclude nothing except that the ponds did not increase clinical mastitis.

This herd was on a lactation treatment study during the summer of 1987. A total of 40 cows were sampled and treated. There were no unusual organisms treated. These 40 cows were not all the cows treated for clinical mastitis during the period the cows had access to the ponds, but the results were encouraging.

Statewide Effects of Ponds on Milk Quality

Regulatory samples for all the herds in Florida were obtained from the Division of Dairy Industry for the year of 1987 and were analysed for the effects of no ponds, man-made ponds and natural ponds. The numbers presented are least squares estimates of Standard Plate Counts (SPC) and Direct Microscope Somatic Cell Counts (DMSCC) for each month. The data are not biased by different numbers of dairies with no ponds, natural, or man-made ponds. The data were analysed by herds whose milking cows had access to ponds as indicated on the survey.

The data indicated that milking cows with access to man-made cooling ponds had lower SPC and DMSCC counts than other herds in the state with no access to ponds and those herds with natural ponds were higher in both categories. From these data it could be argued that people who built man-made ponds were better managers and thus had higher milk quality.

ENZYMES INVOLVED IN DIGESTION

In animal nutrition, proteins, carbohydrates, fats and phosphatids are acted upon by digestive enzymes. A vast amount of work has been done on the various enzymes found in animals, from protozoa to man. Of course the major part of the work has been done on mammals, but there is an increasing number of papers on the enzymes of lower forms. Presumably the number of different enzymes is legion, for there is no reason to suppose that the pepsin of a pig's stomach is identical with the pepsin from the stomach of a sheep, any more than pig hemoglobin is identical with sheep hemoglobin.

However, it is doubtful if at the present stage of progress in enzymology it would be profitable to attempt a separation of the various individual pepsins or other enzymes. So much more important work now needs to be done.

It is scarcely necessary to consider separately the enzymes of each group of animals. For the general physiologist, primarily concerned with problems of broad over-all importance, there is relatively little interest in the knowledge of which organisms possess which enzymes. On the other hand, the subject is of direct concern to the comparative physiologist. Numerous references to the extensive literature are given by Prosser in his very useful book on comparative physiology.

In the animal organism proteins are digested by the action of hydrolases which attack either the intact protein molecule or intermediate products of hydrolysis. As a final result of the digestion amino acids are produced. The classification of these enzymes is difficult. Certain enzymes, the proteinases, can digest intact proteins; others, the peptidases, can digest only peptides.

However, this classification breaks down, for proteinases can also act as peptidases. Possibly a better distinction is between endopeptidases, or true proteinases, and exopeptidases.

The latter can act only on relatively simple compounds which possess one or more free terminal groups such as α-amino or α-carboxyl groups or both; on the other hand the endopeptidases can act on peptide bonds in any part of the protein molecule.A useful, though artificial and not absolutely correct classification can be made on the basis of the optimal *p*H of the enzyme. This classification is not exact, for optimal *p*H may to some extent vary with the substrate. Animal digestive proteinases are of three main types:

- Pepsins. These act at strongly acid *p*H—the optimum is usually around 1.5 to 2.
- Trypsins act in neutral or slightly alkaline solutions—the optimal *p*H is on the alkaline side of neutrality. In higher animals the pancreas secretes both trypsin and chymotrypsin; these are distinct enzymes-chymotrypsin has a strong milk-clotting action.
- Cathepsins act most effectively at the isoelectric points of proteins—hence their optimal *p*H is between that of pepsin and that of trypsin. Cathepsins are found in various tissue cells of animals; they have

properties similar to those of the papainase enzymes of certain plants.

In addition to these three groups of enzymes, rennin, the enzyme which curds milk, is now generally regarded as a separate proteolytic enzyme in spite of the fact that Northrop's crystalline pepsin shows rennin activity. Indeed rennin has been prepared in crystalline form; the pure crystalline enzyme shows definite proteolytic activity. Although rennin is an enzyme concerned with the digestion of milk, it is found in some fishes which obviously do not have milk in their diet. However, there does not appear to be any rennin in the gastric juice of an adult man.

Apparently, pepsin only occurs in the phylum Chordata. Lower chordates such as Amphioxus do not secrete pepsin. Moreover, the various species of fishes which are without stomachs also lack pepsin. When present, pepsin is secreted by glands of the stomach. In frogs and some other amphibia, it is also secreted by glands located in the esophagus. Pepsin does not appear to be present in insects, and the older reports of pepsin in protozoa are doubtful.

Trypsin is found not only in vertebrates, but in various invertebrates as well. Mouton described the extraction of trypsin from amebae, but his observations, made long years before the discovery of cathepsin, really demonstrate clearly the presence of cathepsin in amebae. (Mouton's enzyme was unable to digest fibrin in solutions just alkaline enough to turn phenolphthalein red; in slightly more acid solutions, it was effective).

Oppenheimer believes that many of the trypsins described for invertebrates are really cathepsins. However, true trypsins have been described for mollusks, crustacea and insects. In vertebrates, trypsin is secreted in the active form by the pancreas. As previously noted, the inactive proenzyme trypsinogen is converted into trypsin, partly as the result of the action of enterokinase, a substance secreted by the intestine.

Cathepsin occurs in protozoa, and probably all proteolytic enzymes of protozoa belong to the cathepsin group. Krijgsman found that Trypanosoma evansi contained no pepsin or trypsin, but only a cathepsin with an optimum *p*H somewhere between 3.5 and 4.9.

The fact that ameba contains cathepsin has already been noted. According to Sawano, the proteinase of paramecium is a cathepsin, and Lawrie's studies indicate the presence of cathepsin in the ciliate, Glaucoma piriformis. Inasmuch as digestion in protozoa is primarily intracellufar, it is readily understandable that the proteinases of protozoan cells are similar to the proteinases found in the tissue cells of higher forms.

One might also expect that in lower forms with predominantly intracellular digestion the proteinases would also be cathepsins. Bodansky and Rose studied the proteinases of coelenterates, and they found two types of enzymes. For one, the optimum *p*H was slightly above neutrality—for the other it was between 3.0 and 3.5 Whether to call either or both of these enzymes cathepsins is largely a matter of definition.

Some proteins are not attacked by ordinary digestive enzymes. Thus keratin, the protein which constitutes the bulk of skin, hair, wool and feathers, is not digested by pepsin or trypsin. However, if keratin is subjected to reducing agents, so that the S-S groups it contains are changed to SH groups, it then becomes susceptible to attack by pepsin and trypsin. This is what happens in the intestine of the clothes moth. Keratin is first reduced and is then digested by a proteinase which acts in a strongly alkaline solution. This proteinase has a *p*H optimum of 9.6-10.2.

The enzymes which hydrolyze proteins (pepsin, trypsin, cathepsin) do not carry digestion very far. The ultimate breakdown to individual amino acids is carried on by the peptidases (or if one prefers, the exopeptidases). There are many such enzymes; in higher animals they are secreted both by the pancreas and by the intestinal mucosa. Numerous peptidases have been described for various organisms from ameba to man.

Some of them appear to have very definite specificity toward certain types of amino acids, but our knowledge of the specificity of these enzymes is limited by the fact that very few of them have been prepared in anything like pure form. Nor is classification by optimal pH possible, for most of them have an optimal pH in the neighbourhood of 7.5-8.5.

One can distinguish between polypeptidases, which act on polypeptides, and dipeptidases which act on dipeptides (although it is not certain that some of the dipeptidases do not also have an effect on polypeptides).

One can also distinguish between carboxypeptidases which act on free carboxyl groups and aminopeptidases which act on free amino groups. An enzyme may be characterized as a carboxypolypeptidase, etc. Details remain to be worked out for almost all types of organisms, and in 1942 Johnson and Berger complained: "It is unfortunate that too few current papers take sufficient cognizance of our complete ignorance of the enzymatic constitution of most peptidase systems."

Fats are digested by lipases, which, in the case of mammals, are secreted by the stomach, intestine and, perhaps, also by the salivary glands. In the intestine the activity of lipase is accelerated by bile, and also by the presence of calcium salts. The digestion of lecithin is not clearly understood. Ordinary lipases may play a part, as well as phosphatases (enzymes which hydrolyze esters of phosphoric acid).

Lecithin may be partly digested by lecithinases, but these have been studied more in bee and snake venom than in intestinal contents. According to Režek, the hepatopancreatic juice of the snail (Helix pomatia) splits lecithin completely into fatty acids, glycerol, phosphate and choline. Apparently various enzymes are involved. Various steps are involved in the complete breakdown of lecithin.

There are a number of different types of enzymes which digest carbohydrates. Although cellulose forms a large fraction of the food of those animals which eat plants, relatively few animals possess cellulosedigesting enzymes. The older literature is thoroughly reviewed by Yonge. A number of workers have described cellulases in crabs, in snails, in clams, and in various other mollusks including Limax, Teredo, etc. Some wood-boring insects possess cellulase, others do not. In most cases, however, digestion of cellulose is accomplished by the bacteria and protozoa present as symbionts in stomach or intestine.

Thus grazing animals such as the cow, sheep, and horse, possess no cellulase of their own, and are dependent on symbiotic organisms. These are present in vast numbers, and there are many species both of bacteria and protozoa.

The latter seem to share with the bacteria the task of digesting cellulose. In spite of the fact that so many animals are covered by chitin, very few organisms possess a chitinase. (Chitin is a complicated carbohydrate similar to cellulose; it is an amino-polysaccharide, and is found in the cell walls of fungi as well as in the integuments of arthropods and other animals).

The literature on chitinase is also reviewed by Yonge. The enzyme is present in the edible snail (Helix pomatia). It is also secreted by the beetle Platyderma tricuspis which feeds on fungi, and by the larva of an ectoparasitic hymenopteran Pseudagenia carbonaria which feeds on spiders. Starch-digesting enzymes are found generally in all types of animals. So, too, are enzymes which convert double sugars into simple sugars.

As a result of the action of digestive enzymes, protein food is broken down into amino acids, fat into fatty acid and glycerol, and carbohydrates are split into simple sugars. In most animals this digestion takes place in the alimentary canal, and the products of digestion are then absorbed through the wall of this canal.

Chapter 5

Intake of Food

PROTOZOA

Feeding habits of protozoa vary widely. Some contain chlorophyll and behave essentially like plants. Thus the various species of Euglena are able to manufacture starch from simple inorganic materials. These forms and also ciliates without mouths, such as the Opalinidae, are able to absorb food materials through their surfaces. The taking up of dissolved substances by Opalina ranarum, the common parasite of the rectum of the frog, was studied by Kedrowsky. In all of the sporozoa, moreover, food enters the organism through the body wall. These forms are parasites and live on the organic materials provided by their hosts.

Most ciliates and flagellates have definite mouths, but it does not follow that all of the food taken up by these organisms passes through the mouth opening. In flagellates, especially, it seems probable that much of the food enters the cell by diffusion through the body wall. Ciliates like paramecium sweep food particles into their gullet by the beat of their cilia. Although various useless materials may enter the gullet and pass into the cell, there is apparently some power of discrimination.

Thus Dembowski finds that paramecium will cast out of the gullet particles of sulfur, glass, or chalk, which may find their way there, whereas particles of starch or egg yolk are taken into the cell. Bozler also studied the selection of food by paramecium. Smaller particles were taken in indiscriminately, but there was very definite evidence of selection between different types of larger particles. Other types of ciliates also show discrimination in their choice of food. Thus, according to Schaeffer, Stentor rejects particles of carmine, glass, and sulfur, and ingests organisms like Euglena and Phacus. The selection is apparently due to mechanical factors, for Schaeffer was unable to make the Phacus unpalatable by soaking it in various chemicals. The ciliate, Dileptus, exercises selection in its choice of food, as does the parasitic ciliate, Balantidium coli. Some infusoria (the suctoria) suck up food through their tentacles.

Various authors have studied the feeding habits of amebae and similar forms. There is a wide variation among different types of amebae. According to Schaeffer, the ameba exercises some choice in its selection of food. It may even learn by experience to reject such things as carmine particles. In taking in food, ameba sends out pseudopodia which surround the object to be ingested in a so-called "food cup". The shape of these food cups varies with the type of material to be ingested, and living organisms are generally surrounded by wider cups.

Small particles may be taken into the body of the ameba without the formation of any food cups, the food passing directly into the protoplasm without any activity on the part of the ameba. In some species of amebae the food cups may be rather elaborate and may simulate organelles.

Thus in Amoeba vespertilio Ivaniè described (in fixed preparations) the presence of mouth-like structures provided with a ring of more densely staining material. Ivaniè believes this ring to be a gel—no doubt it is part of the cortical gel known to be present in ameba. When amebae are placed in hypertonic solutions, they form tubes which ingest water.

A similar phenomenon has been described for various types of cells in tissue culture by Lewis. He calls the ingestion of water by cells "pinocytosis." It is possible to make quantitative studies of the rate at which food particles are ingested by ameba. Mast and Fennell studied the rate at which Chilomonas was ingested by Amoeba proteus and Amoeba dubia. When the concentration of Chilomonas was doubled, the number of flagellates ingested was only slightly increased.

Mast and Fennell also studied the effect of various temperatures and hydrogen ion concentrations on the rate of ingestion.

The ingestion of food by ameba is essentially the same process as that which occurs when leukocytes or other ameboid cells take up foreign particles.

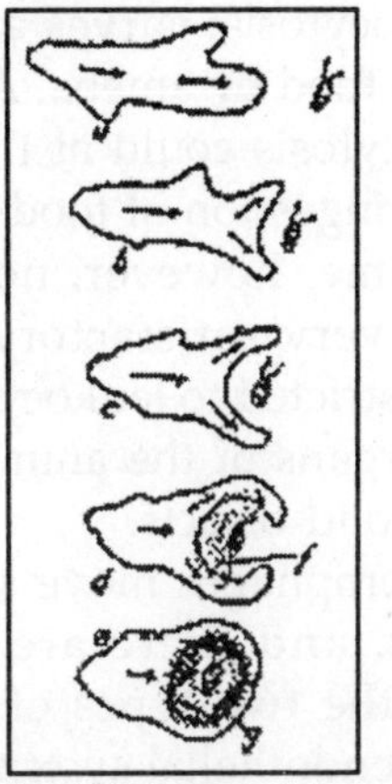

Fig. Food Cups of Ameba

Such phagocytosis occurs in practically all metazoa, and is of great importance both in the normal life of animals and in the resistance of animals to disease. Phagocytosis plays an essential role in metamorphosis and, both in insects and in amphibia, larval organs are destroyed bit by bit by the ameboid cells which ingest them. Because of its importance as a protection against disease, phagocytosis has been very widely studied by bacteriologists, pathologists, and students of medical sciences generally.

Just as amebae select their food, so do leukocytes exercise a selection of the particles they ingest. Thus Fenn showed that leukocytes ingested carbon particles four times as rapidly as quartz particles.

The rate of ingestion of bacteria by leukocytes has often been studied under a variety of conditions. Bacteriologists have tried to obtain quantitative data by determining either the number of bacilli taken up per leukocyte or the percentage of leukocytes participating. Important factors (not always properly considered) are the absolute and relative concentrations of bacteria and leukocytes.

Phagocytosis increases with increase in the number of bacteria, but the rate of ingestion does not keep pace with the increasing possibilities for collision between bacteria and leukocytes. Various authors have studied the effect of temperature on phagocytosis; curves are obtained like those found for ingestion of food by ameba. Any explanation of the mechanism of phagocytosis could at the same time serve as an explanation of the ingestion of food by ameba.

At the present time, however, none of the theories of phagocytosis seems very satisfactory. In higher animals, phagocytosis is not restricted to leukocytes. Scattered through various tissues and organs of the animal are numerous cells capable of ingesting solid objects.

Of these, the macrophages move somewhat more freely than the histiocytes, and there are also morphological differences between the two types of cells. Together, they constitute the reticuloendothelial system. Many authors have described phagocytic cells in embryonic tissues.

METAZOA

Most many-celled animals take solid food into some cavity within their bodies, in which it is dissolved or digested. This is not always the case, however. There are quite a few organisms which are able to digest their food outside of their bodies. Thus a starfish or sea urchin may pour digestive juices out over some fish or other large animal which it has captured. It is sometimes possible for an animal to eat other animals larger than itself. When a spider captures a fly, it pours digestive juices into the body of its prey and then sucks out the dissolved interior.

Many parasitic animals require no digestive apparatus. In the case of animals such as tapeworms, there is no need for the ingestion of solid food, and the dissolved food substances in the alimentary canal of the host pass directly through the body wall of the animal. Other worms derive food in the same manner from the blood of their hosts. Some authors have tried to classify animals on the basis of their methods of ingestion. Thus there have been distinctions between whirlers (ciliary feeders), snarers, scrapers and suckers.

The most elaborate classification is that of Yonge and his review should be consulted for references to the literature on invertebrates.

Vertebrate ingestion is discussed by Böker. On the basis of their ingestion habits, Yonge divides animals into three major groups: those that ingest small particles, those that take in larger particles or masses, and those that suck or absorb food. Small particles are often pulled into the body by ciliary action.

This occurs not only in the protozoa, but also in practically every other phylum; except the arthropoda. Ciliary mechanisms for feeding are most highly developed in the mollusca. Tentacles and setae are also used in obtaining small particles.

The animals that feed on large particles or masses have mechanisms for swallowing inactive food, for seizing prey, or for scraping and boring. The usual form of ingestion in metazoa involves some sort of passage into the mouth. Jaws and teeth of various types may aid in the capture and in the subdivision of the food. The earthworm has a pharynx which is pulled on by muscles attached to its outer wall so that it can behave as a suction pump. Marine (polychaete) worms also may have sucking pumps. Thus in Autolytus the proventriculus, by its pulsations (120 per minute), produces a strong inward current of water and food.

Sucking pumps also occur in insects of various orders, and these pumps may have a rather complicated structure. Even in the case of those animals which ingest food through the mouth there is sometimes a possibility that food may be taken in through the body wall.

Development of this idea is due especially to Pütter who thought that many types of organisms receive nutriment from organic materials dissolved in sea-water or fresh water. Pütter believed that chlorophyll-containing plants synthesized more organic substances than they themselves could use, and as a result an excess was then given off to the surrounding medium from which it could be taken up by animals.

Some animals were thought to take up food directly through the body wall. In the case of fishes Pütter thought that the food entered through the epithelium surrounding the gills. The ideas of Pütter were sharply challenged by Lipschütz. Since then there have been a number of papers on the subject. The literature up to 1931 is reviewed by Krogh. Although he notes cases of absorption of organic nutrient substances when these are present in relatively high concentration, Krogh concludes that among metazoa, "there is no convincing evidence that any animal takes up dissolved organic substances from natural water in any significant amount."

Matheson and Hinman have urged that mosquito larvae do take up organic substances from natural waters and Hinman has reared such larvae in sterile solutions in which their nutrition depended solely on organic materials in solution.

Hinman believes that "many aquatic animals probably utilize organic material in true solution," but direct evidence for animals in their natural environment is lacking. In the case of the mosquito larva, there is no knowledge as to what proportion of organic material is absorbed through the skin and what proportion through the intestine. Mosquito larvae absorb chloride through the cuticle of the gills. Various other freshwater animals also absorb ions through the integument.

DIGESTION—ENZYMES

Whenever an organism consumes food in the solid state, this must be brought into solution before it can be utilized. It is necessary, therefore, that such solid food be digested. In some organisms, as pointed out in the last chapter, digestion may occur outside of the body, and this may constitute an important adaptation for those animals which are in the habit of eating animals larger than themselves. Ordinarily, however, food is taken into the body and digested there.

Digestion may occur in cavities of special organs such as the stomach or intestine, or it may occur within the protoplasm of cells. The latter type of digestion obviously takes place in protozoa. In organisms such as paramecium or ameba the ingested food is enclosed in a food vacuole, which serves the same purpose as the stomach or intestine of a complicated metazoan.

Within the food vacuole the solid food particle is brought into solution. It must not be thought, however, that intracellular digestion is confined to protozoa. In sponges, coelenterates, and flatworms, much of the solid food taken into the body is ingested by ameboid cells lining the walls of the digestive tract, and digestion takes place within the protoplasm of these cells rather than in the lumen of the digestive tube or cavity.

Moreover, in some animals rather higher in the evolutionary scale, there is also a certain amount of intracellular digestion. In spiders and in arachnids generally, the digestion of protein, begun either outside the animal or in the digestive tract, is finally completed within the cells lining the tract. In clams and other lamellibranchs, the digestion of protein and fat has been thought to occur exclusively within cells.

Both in lamellibranchs and echinoderms, amebocyte cells play an important role. These phagocytic cells even enter into the lumen of the stomach or intestine, ingest particles of food there and then carry these food particles back into the tissues and digest them there. Such intracellular digestion, however, is a primitive character and it does not occur to any extent in higher animals such as the insects and vertebrates.

In the conversion of solid food to a state of solution, enzymes play the leading role. It will be necessary, therefore, to consider the subject of enzymes and the nature of enzyme action. It should be strongly emphasized, however, that enzymes are not concerned only with digestion, but that they are essential factors in all of the chemical activities of the organism.

Our discussion at this point will be somewhat parenthetical. It will also be brief; indeed, it would scarcely be possible to present an up to date summary of enzyme chemistry in the space of a single chapter. The subject has grown so rapidly that it has become a science by itself (enzymology), and the modern books on enzymes are heavy with information.

DEFINITION OF ENZYMES

The living machine is essentially a chemical engine, dependent for its growth, maintenance and energy on chemical reactions. These reactions are controlled by catalysts. One of the most striking achievements of modern biology and biological chemistry is the isolation of more and more of these catalysts so important for the vital process. Once isolated, it is possible to study their behaviour in test tubes or other suitable containers. In other words, one can detach from the living material certain non-living substances capable of causing or promoting the complex chemical transformations which constitute a major part of the mystery of protoplasm.

Originally, enzymes were defined as "catalytic substances produced by living cells." With increase in knowledge and understanding, it has become necessary to restrict this definition. The original concept was that enzymes were not only produced by living things but were peculiar to them. There are some rather simple substances in protoplasm which can act catalytically; so for example, ascorbic acid, glutathione (a tripeptid important in respiration, or as a matter of fact, even the hydrogen ion.

The term enzyme was never meant to apply to these. Hence in modern usage an enzyme is defined as a "catalyst of biological origin, possessing a high molecular weight." In the light of what we now know, it is generally believed that all enzymes are proteins.

Inasmuch as enzymes are catalysts, it is important to remember that they hasten rather than initiate reactions, and that when they act on reversible reactions they accelerate the progress of the reaction in either direction.

Thus the same enzyme may synthesize as well as break down a complex chemical compound, and the very enzyme which digests proteins or fats or carbohydrates also serves, under the proper conditions, to build up these compounds from their breakdown products.

The substance acted upon by an enzyme is called its substrate. In designating a particular enzyme, the suffix -ase is added to the name of the substrate. However, older names, such as pepsin or rennin, are retained. The suffix -lytic is used to denote enzyme activity on classes of substances. Thus, proteolytic enzymes act on proteins, and lipolytic enzymes act on fats or lipids.

CLASSIFICATION OF ENZYMES

For many years it was the custom to classify enzymes solely by the reactions which they catalyzed. As already noted there is within living systems an amazing variety of chemical reactions and these many diverse reactions are specifically catalyzed by individual enzymes. However, as the chemical composition of enzymes becomes better understood, there is an increasing tendency to classify enzymes not merely in terms of what they do but also in terms of their true chemical composition.

In the first place it is possible to characterize a large group of enzymes as hydrolyzing enzymes or *hydrolases*. The hydrolases are concerned primarily with the breaking down and the building up of proteins, carbohydrates and lipids. It is a characteristic of these essential organic materials that their large molecules are built up by a progressive subtraction of water molecules. When amino acids unite, the acid group of one amino acid combines with the basic amino group of a second, and as in typical acid-base reactions, water is split off.

In this way protein molecules are built up, and if they are to be broken down to amino acids again, water molecules must be added.Similarly, when an alcohol unites with an organic acid to form an ester, water is split off; and in the reverse process by the addition of water an ester may be converted into an acid and an alcohol.

Thus by the addition of water, a fat may be hydrolyzed to fatty acid and the trivalent alcohol glycerol. When two molecules of a simple sugar such as dextrose unite to form a disaccharide, then also water is split off, and with the addition of water the double sugar may be broken into simple sugars again. Similarly starch is formed from sugar by a series of reactions in which water is removed. All of these reactions in which water is removed or added are catalyzed by the hydrolases.

Proteolytic enzymes break down proteins to polypeptids and amino acids (or build up proteins from such compounds), lipases split fats into fatty acids and glycerol (or synthesize the fats from the fatty acids and glycerol), amylase acts on starch, sucrase (often called invertase) acts on sucrose, and there are other specific enzymes which act on the various isomers of sucrose.

The hydrolases also include the *lecithinases* which split off fatty acids from the rest of the lecithin molecule, and the *phosphatases*. The latter split off phosphate from organic combination. Thus one phosphatase breaks glycerol phosphate into glycerol and phosphate; another group of phosphatase enzymes, the nucleotidases, break nucleotides into nucleosides and phosphoric acid. Most hydrolases are concerned primarily with the breakdown of large molecules as it occurs in digestion and with the synthesis of the breakdown products to form large molecules again in the cells. Such reactions are of primary importance in nutrition, but they yield little energy.

In addition to the hydrolases concerned primarily with nutrition, there are some hydrolases which play a part in other vital phenomena. Thus the enzyme cholinesterase, which hydrolyzes acetylcholine into choline and acetic acid, is believed to have great importance for nervous function. Another hydrolase enzyme which has excited great interest is hyaluronidase.

This breaks down hyaluronic acid, a polysaccharide that tends to hold cells together. The enzyme is found in various bacteria; also in testis extracts. When skin is treated with hyaluronidase, the cells separate sufficiently to permit the spreading between them of substances like India ink.

Heparin, like hyaluronic acid, is a mucopolysaccharide. There is apparently a heparinase which breaks up heparin or at least destroys its activity. Mention has already been made of the enzymes which specifically attack ribonucleic acid and desoxyribonucleic acid.

These likewise are hydrolases. The enzymes which cause coagulation or clotting, thrombin and rennin, may also be hydrolases. Thrombin causes clotting of vertebrate blood and rennin clots milk. They may both produce this effect as a result of proteolytic action. Similar to the hydrolases are the *phosphorylases*.

These enzymes, which should not be confused with the phosphatases, add phosphoric acid instead of water. They are very important in carbohydrate metabolism and will be discussed later in this connection. Enzymes which act on inorganic pyrophosphates and hydrolyze them to orthophosphates are called *pyrophosphatases*. (This is a reaction which occurs even without the aid of enzymes). Pyrophosphatases are found in many different types of living material. In addition to the hydrolases there are various other categories of enzymes.

There are enzymes which transfer a radical or group from one substance, the donor, to another, the acceptor. These have been called *transferases*, and they include the transaminases, which transfer amino groups; and the phosphokinases, which transfer phosphate groups. An interesting enzyme found especially in the red blood cells of higher animals, is carbonic anhydrase. This catalyzes the reaction in which carbonic acid is split into carbon dioxide and water.

Actually, the reaction is speeded in both directions, so that carbonic acid can be formed more rapidly as well as broken down. The reaction needs to be speeded in order to hasten the gain or loss of carbon dioxide from the blood. In older books on enzyme chemistry, all enzymes which were not hydrolases were called desmolases. The term was originally coined for enzymes that break the carbon chain, that is to say, the C—C bond, and it is perhaps best restricted to these enzymes. Among the true desmolases are the carboxylases.

The action of carboxylase enzymes was first studied in investigations on the alcoholic fermentation produced by yeast. In this system, a carboxylase enzyme acts on pyruvic acid, an acid of great importance in carbohydrate metabolism. Pyruvic acid (CH_3 COCOOH), an alpha keto acid which can be considered as a derivative of propionic acid, is split into acetaldehyde and carbon dioxide by carboxylase. Thus:

$$CH_3\,COCOOH \rightarrow CO_2 + CH_3\,CHO.$$

In addition to the carboxylase enzymes which act on alpha keto acids, there are also carboxylases which act on alpha amino acids, converting them to amines: Thus:

$$R.CHNH_2\,COOH \rightarrow CO_2 + R.CH_2\,NH_2.$$

Enzymes of this sort are known to act on tryptophan, tyrosine and histidine. Carboxylase converts histidine into the important humoural substance, histamine. Many enzymes promote oxidative processes. These have been much studied and there is increasing information as to their chemical nature, so that their classification is in part based on what the enzymes are rather than on what they do. In this section no attempt has been made to present a complete classification of known enzymes. Such a classification would cover many pages. All that has been done here is to indicate the broad outlines of modern classification schemes and to indicate the names and activities of a few important enzymes.

PREPARATION OF ENZYMES

This may vary widely. Enzymes occur in some secretions such as saliva or gastric juice. Usually they are prepared from tissues which are chopped up and extracted with water, salt, acid, or basic solutions, or glycerol. Some enzymes (occasionally called "endo-enzymes") resist separation from cells, and special methods must be used to extract them.

The cells are frozen and ground up, or they may be subjected to high pressure with a hydraulic press. Purification of crude extracts may be accomplished by dialysis, by selective adsorption on various types of adsorbents or by crystallization.

CHEMICAL CONSTITUTION OF ENZYMES

Although some earlier investigators reported preparation of enzymes which gave no protein tests, and although the famous Willsätter school maintained for years that enzymes might be bound to protein but were not in themselves protein, it now seems clear that enzymes are actually protein. Sumner, in 1926, prepared a very potent crystalline preparation of urease. The octahedral crystals are apparently a globulin.

Recrystallization does not change the potency of the preparation and the activity seems to be due to the globulin rather than to some substance bound or adsorbed to it. In 1930, Northrop obtained a crystalline preparation of pepsin. The crystals are very similar to the urease crystals obtained by Sumner, and differ only in having a hexagonal instead of an octagonal base. Northrop and Kunitz have also isolated crystalline trypsin.

More recently, crystalline protein preparations of various other enzymes have been obtained. In these crystalline preparations of enzymes there is evidence that the activity is really due to the protein and not to an impurity or to an adsorbed compound, for denaturation of the protein results in a loss of activity of the enzyme, and recovery from denaturation carries with it a proportional recovery in enzyme activity.

PHYSICAL PROPERTIES OF ENZYMES

In general, enzymes behave as colloids or as substances of high molecular weight. They diffuse slowly and do not pass through collodion or similar membranes. By studying the osmotic pressure, the rate of diffusion, and by chemical methods, Northrop determined the molecular weight of his crystalline pepsin. The various methods gave values which ranged from 33,000 to 38,000. Like proteins, enzymes may be salted out. They also migrate in an electric current. One of the most striking properties of enzymes is the fact that their activity is destroyed by heat.

Indeed, the presence of enzymes is often demonstrated by this sensitivity to high temperature. Thus if one desires to show that a reaction is catalyzed by enzymes, one may heat the reacting system and note if the reaction ceases. However, the destructive effect of heat is not always irreversible as was formerly thought. Thus, Northrop showed that crystalline trypsin inactivated by heat could recover its activity if cooled again. The purer the crystalline preparation, the less sensitive to heat it became.

FOOD REQUIREMENTS

VITAMINS

The fact that human beings were subject to various diseases as a result of deficient diet has been long known. Even the early Egyptians knew that night blindness could be cured by eating the livers of domestic animals. Scurvy occurred among the crusaders of the thirteenth century. In 1520 the Austrian surgeon, Kramer, pointed out that scurvy could be cured by oranges, limes, or lemons, and in 1795 lemon juice was introduced in the English navy as a preventive of the disease.

In 1882 the Japanese admiral Takaki (or Takagi) was able to cause a marked decrease in the number of cases of beriberi in the Japanese navy by improving the diet of the sailors. Vitamin research really began with the discovery that animals could be used in experimentation with deficiency diseases.

The first clear-cut experimental work is that of Eijkman. He found that if chickens were fed polished rice they developed a disease, called polyneuritis, very similar to beriberi. This disease could then be cured by feeding the birds the polishings which had been removed from the rice. Following this important experiment, progress was for a time slow, and most workers in the field of nutrition did not realise its significance.

Hopkins was one of the exceptions. As far back as 1906 he noted that no animal could live on a diet of pure protein, fat, and carbohydrate, and that obscure dietary factors were responsible for such diseases as scurvy and rickets. It was not until 1915 that it was realised that more than one type of vitamin was involved in nutrition.

In that year McCollum and Davis showed that two types of unknown substances were necessary for the normal nutrition of the rat. At first these were called "fat-soluble A" and "water-soluble B," but it was not long before the term vitamin was accepted throughout the world, and fatsoluble A and water-soluble B became known as vitamin A and vitamin B. Soon there came to be a growing alphabet of vitamins. Then followed a period in which vitamin B began to be subdivided into B1, B2, etc.

At the present time it is customary to speak of the vitamins A or the vitamins of the B complex. As we come to know the true chemical composition of the individual vitamins, there is an increasing tendency to call them by their correct chemical names instead of by letters. Thus vitamin C is now commonly referred to as ascorbic acid.

There is no clear-cut definition of vitamins, no sure way of deciding whether a substance important in nutrition is to be regarded as a vitamin or not. Vitamins are organic substances capable in small concentration of affecting the health or influencing the rate of growth of organisms. There are numerous substances that act in this way. Some of them are regarded as vitamins, others not, or there may be differences of opinion.

Thus it was noted in the last chapter that certain unsaturated fats were necessary for the growth and well-being of rats. These fats can be regarded as a fat requirement or they can be called vitamin F. Similarly, various insects require cholesterol.

This may be regarded simply as a sterol requirement, or the cholesterol may be considered a vitamin in the same sense that calciferol (vitamin D) is a vitamin. The fact that vitamins are capable of exerting a powerful influence when present in extremely low concentrations is an indication that they act in some way as catalysts.

Certainly this is true of some vitamins. It is by no means an easy task to present an up-to-date summary of our information concerning vitamins.

New papers are being published at a rapid rate, and it is well-nigh impossible for anyone not a specialist in the field to keep abreast of them.

For the student, perhaps the easiest way to become informed is to read some book on the subject or some recent treatise on physiological chemistry. The earlier studies of vitamins were essentially practical. Men suffered from diseases due to vitamin deficiency. It was important to discover which foods contained the necessary vitamins, and in what amount. Lacking a chemical knowledge of the vitamins it was essential to develop biological methods of assay.

Thus, it was found that a certain amount of vitamin was necessary to cure pigeons of polyneuritis or to permit normal growth in rats. On the basis of such studies on rats, pigeons, and guinea-pigs, various types of vitamin units were established, and the vitamin content of all sorts of foods was determined.

To raise colonies of rats and guinea pigs is rather an expensive and time-consuming business. Fortunately, the presence of vitamins and the amount of a certain vitamin contained in a given sample of food can be tested with organisms much easier and less expensive to raise than mammals and birds. Insects require many of the known vitamins and they might be used as test animals. But much more suitable are various bacteria and fungi. They can be raised in test tubes in a minimum of space. Obviously one must first determine the exact vitamin requirements of a given type of bacterium or fungus.

The assay of vitamins by the use of bacteria is now well established practice and there is a large literature in the field. This has been reviewed by Snell. In 1941, Beadle and Tatum by x-ray treatment were able to produce mutant strains of the fungus Neurospora which are highly useful for vitamin assay. The usual type of Neurospora requires only the vitamin biotin, but by the use of x-rays and ultraviolet radiation within a few years mutants were obtained which had a specific requirement for every one of the B vitamins then known with the single exception of folic acid.

Similar mutants have also been obtained in bacteria. The chemical composition of most vitamins is now known. This makes possible direct chemical analysis.

Often such analysis is aided by spectroscopic study. Thus the A vitamins give a blue colour with antimony trichloride and this colour is the basis for spectroscopic tests. Some vitamins give characteristic absorption bands in the ultraviolet. Likewise, some vitamins are fluorescent, and the fluorescent colour can be tested spectroscopically.

Thus in addition to ordinary chemical assay methods, spectroscopic tests are of great value. The subject is discussed at length in volume 1 of György *Vitamin Methods*. In many cases a vitamin may be replaced by other chemical substances related to it chemically. Thus various substances related to ascorbic acid (vitamin C) behave like ascorbic acid in preventing scurvy.

There are many substances with vitamin D or vitamin K activity. The fact that a substance has a chemical structure similar to that of a vitamin does not necessarily indicate that the substance will act like the vitamin. Often, indeed, a substance which resembles a particular vitamin in its chemical composition is an antagonist to it and tends to prevent its action. So, for example, Dicumarol is chemically close to vitamin K; and yet whereas feeding of vitamin K favours blood clotting in higher animals, Dicumarol prevents clotting. Similarly, various analogues of pyridoxine (a B_6 vitamin) act to prevent its action. Other cases of anti-vitamins could be cited.

Studies of antivitamin action may give clues as to the mechanism of the action of the vitamin they antagonize. The subject is also of considerable interest to the theoretical pharmacologist. More and more cases are being discovered in which the harmful action of a particular drug is due to its chemical affinity to some substance important for the vital process. In some cases, the need for a vitamin is lessened by feeding a substance which can be converted into it. In some organisms, tryptophan can be converted into nicotinic acid. Thus, chicks do not require nicotinic acid if fed enough tryptophan. Moreover, isotope studies show that tryptophan can be converted into nicotinic acid.

Another instance in which a necessary vitamin can be replaced by an amino acid has been described for the lactic acid bacteria, Streptococcus faecalis and Lactobacillus casei. Certain strains of these species require vitamin B_6, but they can grow without this vitamin in the presence of *d*alanine (the dextro-rotatory, naturally occurring form of the amino acid). In this instance, however, the amino acid is not the precursor of the vitamin, but apparently the vitamin is necessary in order to produce the essential amino acid.

These cases, interesting in themselves, indicate also that the need of a particular organism for a given vitamin may to some extent depend on the diet. As a result of our knowledge concerning the chemical nature of the vitamins, many of them can be synthesized and are actually being manufactured on a large scale.

We also have much knowledge concerning the specific vitamin requirements of many higher animals, of the isolated roots of green plants, of bacteria, fungi and protozoa. More and more information is becoming available concerning the vitamin needs of various insects. Vitamin studies are relatively simple for terrestrial forms-mammals, birds, insects. Aquatic forms are difficult material.

Perhaps if fish could be grown in sterilized water and kept sterile, correct data on their vitamin requirements could be obtained. With ordinary methods of attack, one can never be certain of the contribution that bacteria and other microorganisms are making to the vitamin content of the food taken in by the fish. Even in terrestrial animals, microorganisms may play a part in providing vitamins.

The bacteria in the intestinal tract can and do manufacture vitamins. In experiments with rats, it is sometimes necessary to prevent the animals from eating their feces, for such fecal material may be a source of vitamins. In many instances a mammal may not need a particular vitamin as food because it is manufactured by the bacteria of its intestine.

By giving the animal antibiotics such as sulfa drugs and thus eliminating or reducing its intestinal bacterial flora, it is possible to investigate the need for a vitamin such as folic acid. The nutritional requirements of various types of animals may depend largely on the parasites or symbionts they harbor.

Many species of insects, especially among the Hemiptera, have a constant association with symbiotic fungi. Even protozoa may contain symbionts or parasites within their single-celled bodies. Most of the work on vitamins has been stimulated by the practical needs of medicine and industry. The physician and the drug manufacturer are not ordinarily interested in the vitamin requirements of lower organisms, unless these organisms can conveniently be used in vitamin assay. Thus, as might be expected, there are vast gaps in our knowledge of the vitamin requirements of many large groups of animals. We know almost nothing of the vitamin needs of metazoan invertebrates other than insects.

And we know very little about the vitamins necessary for fish, amphibia and reptiles. A wide field is open for such study, and many facts of interest may well be discovered. Some vitamins are required by practically all living organisms from bacteria to man; others are needed by only some forms and not others. There has been much interest in the avitaminoses, that is to say the diseases caused by the lack of specific vitamins.

Our knowledge in this field is almost entirely confined to higher animals. Thus we know that when the vitamins D are omitted from the diet, a child will develop rickets and so will a pig or a rabbit, a dog or a chicken. Accordingly, the vitamins D have been called the antirickets vitamins. But animals without bones such as insects or protozoa may also require vitamin D. From a knowledge of the reasons why this is so, we may be able to obtain additional information as to the way that vitamins function.

Much as is known about the vitamins, their chemistry, their occurrence in various foods, the diseases that follow their absence from the diet, there is still almost complete ignorance as to the exact nature of their action. A beginning has been made in the recognition of the fact that some vitamins such as riboflavin, nicotinic acid, and pantothenic acid form essential parts of some enzyme systems.

We know also the end results of many types of vitamin deficiency. But to say that a vitamin is necessary for growth or that it prevents anemia tells us very little as to what it really does or why it is necessary. If we were to know how certain vitamins influence growth or prevent cell deterioration in the epithelial or nervous system, we would have basic information of great importance. Such information would be useful not only in the better understanding of the vitamins but also in the elucidation of various basic vital processes. The final interpretation of vitamin action must lie in an understanding of how the vitamins affect the cell protoplasm.

It is this aspect of the subject that will be stressed in the following discussion of individual vitamins. The general physiologist is not primarily interested in the relation of vitamins to human welfare, the foods a man should eat, or the benefits to be derived from adding this or that vitamin to the diet. Such information can be obtained in books on physiological chemistry, nutrition, or medicine. In discussing the individual vitamins, we shall not proceed in alphabetical order. This order has no other basis than the chronology of discovery. We shall consider first the fat-soluble and then the water-soluble vitamins.

In general, especially in the case of the water-soluble vitamins, true chemical names have tended to supplant letters and subscripts. But in some instances, several chemical substances can and normally do supply a given vitamin need. In these instances the alphabetical name is sometimes more convenient.

VITAMIN A

Vitamin A is a yellow viscous oil. Chemically, it is related to carotene, the common yellow pigment of plants. The generally accepted formula follows:

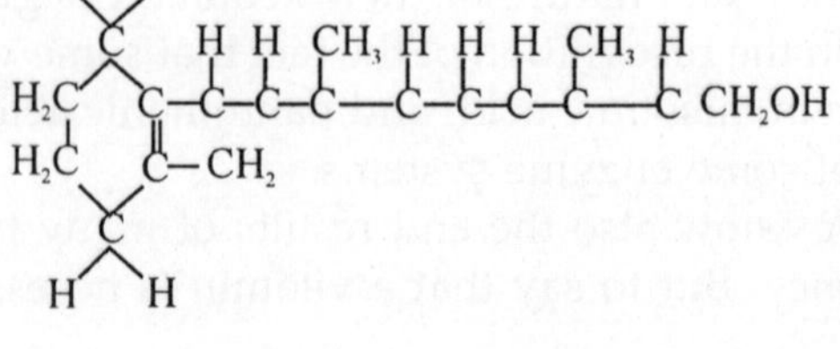

When carotene is eaten, it is converted to vitamin A. Because of this, it is sometimes called a provitamin. Another provitamin of vitamin A is kitol, a dihydric alcohol found in whale liver oil. The conversion of carotene to vitamin A occurs in the intestines of higher animals. The vitamin is then stored in the liver. The livers of polar bears and Arctic foxes are so rich in vitamin A that they are toxic. The toxicity is an example of what is known as hypervitaminosis.

Vitamin A has been synthesized in several different laboratories. There are really two forms of the vitamin. One form (A1) constitutes most of the vitamin content of the livers of salt water fishes. In the livers of fresh water fishes vitamin A_2 occurs. The 2 substances give slightly different colours with antimony trichloride and can be distinguished spectroscopically.

In higher animals, vitamin A is important for growth. When rats are deprived of vitamin A, the cornea of the eye becomes horny. This condition is known as xerophthalmia. It can occur also in children. Not only the corneal epithelium, but other types of epithelia are also adversely affected in vitamin A deficiency; this may lead to lessened resistance to infection. Vitamin A has a relation to vision. In man, vitamin A deficiency causes night blindness, but the effect is not as readily produced as formerly supposed.

In so far as they have been investigated, invertebrates, with the apparent exception of the snail Helix, do not seem to require vitamin A. Aside from its relation to vision, little is known as to the reason why vitamin A is necessary. Its presence in cells can be recognized with a fluorescence microscope because of the fact that in ultraviolet light it gives a green fluorescence.

STEROLS-VITAMIN D

A number of sterols have vitamin D activity, that is to say they prevent the occurrence of rickets in higher animals. The chemistry of the sterols is very complicated, and books on physiological chemistry may be consulted for structural formulae, etc. When the common sterol of fungi, ergosterol, is irradiated with ultraviolet rays, another sterol, calciferol, is produced. This is commonly referred to as vitamin D_2. (There is no D_1, for the original use of the term was a misnomer). D_3 is 7-dehydrocholesterol.

Numerous other antirachitic (rickets-preventing) substances have also been described. Whether or not a given sterol or sterol derivative has antirachitic action can only be determined by actual test. Bacteria do not seem to require sterols.

Some species of protozoan flagellates do; also Entamoeba histolytica, the ameba that causes human dysentery. Insects, in so far as they have been studied, require sterols, but they seem to need cholesterol, a cholesterol derivative, or a plant sterol; calciferol cannot usually be substituted for these. In mammals and birds, vitamin D causes an increase in the deposition of calcium and phosphorus in the bones.

This is apparently due, in part at least, to an increased absorption through the intestine. Using radioactive isotopes, Greenberg found that vitamin D promotes the absorption of calcium and strontium from the digestive tract. Sterols may act in the same way in promoting the passage of calcium through the walls of the insect intestine. Moreover, they may be concerned with the entrance of calcium into cells. This aspect of the subject has never been investigated.

VITAMIN E

A third fat-soluble vitamin or group of vitamins is vitamin E, or perhaps one should say the vitamins E. There are very many substances with vitamin E activity, but the term vitamin E is restricted to the tocopherols. (The word tocopherol is derived from Greek words signifying child birth). There are 4 tocopherols now known: alpha, beta, gamma and delta tocopherol. The formula for alpha tocopherol is:

Alpha Tocopherol

Beta and gamma tocopherol have 1 less methyl group than alpha tocopherol and delta tocopherol has 2 less methyl groups. Vitamin E is necessary for the reproduction of the rat and other mammals and birds. If the male rat is fed a diet lacking in vitamin E, the germ cells in the testis degenerate. (The effect is not evident in the mouse, but is said to occur in the little fish commonly known as the guppy).

In the absence of vitamin E, the female rat cannot give birth to its young; these degenerate and are resorbed. In some cases, muscular atrophy follows lack of vitamin E in the diet. In 1949, Harris wrote: "The mechanism whereby vitamin E functions in the body is still unknown."

A dual role was proposed by Hickman and Harris. In the first place it may act as part of an enzyme system, and to this action its effect in preventing degeneration of muscle may be due. Secondly, the tocopherols exert another, less specific function in that they tend to prevent excessive oxidation. Thus, for example, they tend to preserve vitamin A and they may also prevent the necessary unsaturated fats from being oxidized.

VITAMIN K

Fourth and last of the commonly recognized fat-soluble vitamins are the vitamins K, so named because they are important for blood coagulation (coagulation is spelled with a K in German and in the Scandinavian languages). The K vitamins are close to the E vitamins chemically. Tocopherols are very similar to the naphthoquinones chemically; the K vitamins are actually naphthoquinones. This is a 1,4 naphthoquinone.

The most active form of vitamin K, a form which can be prepared synthetically, is 2-methyl-1,4-naphthoquinone. This is sometimes called menadione. It does not occur naturally nearly as abundantly as vitamin K1, which resembles menadione except that instead of having a methyl group in the 3 position, it has a phytol group.

(Phytol is a 20 carbon mono-unsaturated primary alcohol which can be obtained from the saponification of chlorophyll). Vitamin K_2 also resembles menadione except that in the 3 position it has a difarnesyl group. (Farnesol is similar to phytol; it is a 15 carbon primary alcohol with 3 unsaturated carbon atoms). Several other naturally occurring vitamin K compounds have also been described. They are all either yellow oils or yellow crystalline solids, practically insoluble in water. Vitamin K is manufactured in one form or another by bacteria, which do not ordinarily require it.

The bacteria in the intestines of mammals produce enough so that normally the mammals have no need of it from outside sources. However, when absorption of fatty compounds is interfered with, as in intestinal disease or in liver disease involving interruption of bile flow, the vitamin becomes essential for man and mammals.

Many birds require the vitamin, for their bacteria can not supply them with sufficient quantity. Vitamin K is necessary for the clotting of the blood of higher animals. The reason for this is not clearly understood.

To some extent opinions differ, but most authorities believe that this action of vitamin K is an indirect effect caused by the entrance into the blood of greater amounts of a substance prothrombin, the precursor of thrombin. Thrombin, which is regarded as an enzyme, is of primary importance in blood clotting.

Certainly the function of vitamin K can not be restricted to its effect on the clotting of blood. Obviously, the bacteria and the plants which produce it must have some need for it themselves. Experiments with isolated cells show that vitamin K is a very powerful protoplasmic clotting agent.

This is not surprising in view of the fact that the clotting of protoplasm is in so many respects similar to the clotting of blood. Although menadione is but sparingly soluble in water, solutions of it cause a coagulation of the protoplasm of marine eggs. Menadione also produces clotting in muscle protoplasm.

ASCORBIC ACID (VITAMIN C)

The fat-soluble vitamins are still generally referred to by their alphabetical names. On the other hand, the water-soluble vitamins are now for the most part usually given chemical names which have no relation to the alphabet. This became necessary when what was originally the B vitamin became a whole assortment of heterogeneous substances, distinguished by somewhat variable and uncertain subscripts of the letter B. At present the water-soluble vitamins include the clearly defined ascorbic acid (vitamin C) plus the vitamins of the B complex.

Because of the fact that vitamin C is one substance, clearly defined and understood, we shall consider it first rather than in its alphabetical order. Ascorbic acid is now well understood chemically and it can indeed be synthesized. The L-form is much more potent than the D-form. Ascorbic acid is really an acid hexose sugar.

Its acidity is due to the dissociation of an enolic hydrogen rather than to opening of the lactone ring. Ascorbic acid is a strongly reducing substance, and because of this, it is easy to demonstrate its presence in cells and tissues. A standard method involves the reduction of silver nitrate. Because of its reducing power, there is doubtless a relation of ascorbic acid to the oxidative processes in cells, but there is still no clear understanding as to the exact function or functions of the vitamin in metabolism.

In animals suffering from scurvy there is lack of intercellular material; presumably the cells enclosing the blood vessels are not held together as firmly as they should be and this is doubtless the cause of hemorrhage. Mucopolysaccharides are known to be important constituents of the cementing substances which hold cells together. In view of the fact that mucopolysaccharides are polymers of acid sugars, it is possible that ascorbic acid is involved in the synthesis of such substances.

There is some indirect evidence for this. Normally in wound healing mucopolysaccharides are detectable in the healing tissues and may have considerable importance for the process. However, when the store of ascorbic acid is depleted, these substances do not appear.

THIAMINE (ANEURINE, VITAMIN B1)

The formula for thiamine is $C_{12} H_{18}N_4 OS$. It is prepared synthetically as the chloride hydrochloride. This is a white crystalline compound readily soluble in water. Thiamine is a combination of a pyrimidine ring and a sulfur-containing thiazole ring.

Apparently all sorts of protoplasm need thiamine. Some types of cells can synthesize it, others can not. Higher organisms generally require the intact thiamine molecule; on the other hand some lower organisms are able to synthesize thiamine provided they can obtain both parts of the molecule, pyrimidine and thiazole. Other organisms may require either the pyrimidine fraction or the thiazole fraction, but not both. Many bacteria require neither and can synthesize whatever thiamine they need.

Lack of thiamine in the diet produces beriberi, a paralytic disease of men rather common in Japan and in various other portions of the world.

In birds, lack of thiamine produces polyneuritis, a disease similar to beriberi. In this disease the nerve fibers show degenerative changes. On a fox ranch in Minnesota owned by a Mr. Chastek, it was found that the inclusion of raw fish in the food given the foxes caused a paralysis now known as Chastek paralysis.

Experimentally, it has been shown that a similar paralysis could be produced in chicks, cats, pigeons and rats. The effect is due to the fact that some species of fish contain in their tissues an enzyme, thiaminase, which destroys thiamine.

Thiaminase is found not only in fish but also in clams. In addition to its effect on the nervous system, lack of thiamine may also cause retardation of growth. Thiamine is important for the utilization of carbohydrate. Combined with 2 phosphate groups as diphosphothiamine, it forms part of an enzyme system that acts on pyruvic acid, a keto acid of great importance in intermediary metabolism. When animals are deprived of thiamine, their tissues are unable to oxidize pyruvic acid.

According to von Muralt, thiamine is also important in the transmission of the nerve impulse. When isolated nerves are stimulated, they are believed to release thiamine to the surrounding solution.

RIBOFLAVIN

Riboflavin is an orange crystalline substance, sparingly soluble in water, and markedly sensitive to light. The riboflavin molecule is complex. For an understanding of its structure and its synthesis as first accomplished by Karrer and his associates in 1935, the larger textbooks of organic chemistry may be consulted. The three rings represent an isoalloxazine constituent and this is combined with ribose, the pentose sugar also present in ribonucleic acids. Although riboflavin is apparently a constituent of all or almost all types of protoplasm, many lower organisms do not require it in their food.

Yeasts and molds have no need for it. Bacteria, in general, can get along without it. However, riboflavin serves as a growth factor for the tetanus bacillus, for lactic acid and propionic acid bacteria, and it is a requirement for Erysipelothrix rhusiopathiae and Listerella monocytogenes. The protozoan ciliates that have been investigated require riboflavin. So, too, do almost all the insects that have been studied. Riboflavin is essential for the growth and normal health of the rat, mouse, dog, pig, man, chicken and turkey. In riboflavin deficiency, various symptoms of ill health may appear.

These include redness and roughness of the cornea of the eye, soreness at the corners of the lips, redness and soreness of the tongue, etc. At one time, it was thought that "ariboflavinosis," the disease caused by riboflavin deficiency, was extremely common. But the symptoms mentioned above are not always due to lack of riboflavin. The presence of riboflavin in cells may be detected by fluorescence microscopy. Riboflavin is important for cell respiration. It and other alloxazine compounds combine with phosphoric acid and protein to form enzyme systems known as flavoproteins.

NICOTINIC ACID (NIACIN)

Nicotinic acid and pyridoxine (to be considered next) are closely related substances. Both are derivatives of pyridine. This substance is a weak base, the molecule of which contains five CH-groups and a nitrogen atom arranged in a six-membered ring. As is customary with ring compounds, the various positions on the ring have numbers. In pyridine, nitrogen occupies the number 1 position. Nicotinic acid is 3-pyridine carboxylic acid. The structural formulas of the acid and its amide are given below:

The amide is more important than the acid itself, for some organisms require it and cannot manufacture it from the acid. Both acid and amide are white crystalline powders, without odor. According to Stephenson, "Nicotinic acid is one of the commonest growth factors needed by bacteria; in other words the ability to synthesize it is one most frequently lost. This may be due to its very wide distribution in soil and natural media." It is an essential requirement for those ciliate protozoa that have been studied (Colpoda and Tetrahymena) and also for most of the insects that have been investigated. Nicotinic acid is essential for the dog, the pig, and for man. However, if enough tryptophan is added to the diet, neither dog nor pig requites it.

As noted previously, tryptophan is a precursor of nicotinic acid, and the tissues can synthesize the vitamin from tryptophan. From a lack of nicotinic acid, dogs become afflicted with what is known as blacktongue disease. In man, nicotinic acid deficiency is the primary cause of pellagra.

This serious disease, common enough in the United States, involves the lining of the digestive tract, the skin, and the nervous system. Nicotinic acid amide enters into the composition of two substances of great importance for the oxidative metabolism of the cell. These are coenzyme I and coenzyme II. Presumably nicotinic acid deficiency is associated with a lack in the tissues of coenzymes I and II. But what connection there is between such a lack and the manifold disturbances present in a disease like pellagra is not at all clear.

PYRIDOXINE (VITAMIN B6)

The term vitamin B6 is now properly used, according to a decision of the American Society of Biological Chemists, to designate three naturally occurring substances: pyridoxine, pyridoxal and pyridoxamine. Pyridoxal is the aldehyde derivative of pyridoxine and has a formyl group instead of a hydroxymethyl group at the 4 position. Pyridoxamine is an amino derivative and has an amino methyl instead of a hydroxymethyl group at the 4 position. The three forms of the vitamin are almost equally effective for higher animals, but some bacteria have specific requirements for one form or the other. Vitamin B_6 is needed by many bacteria, by some protozoa, and by most of the insects whose vitamin requirements have been investigated.

When rats are deprived of vitamin B_6, they develop a painful inflammation of the skin. This is especially evident in the skin of the extremities—the paws, snout, and ears. The ailment is sometimes called rat acrodynia because of a supposed similarity to a human ailment in which the nerves and skin of the hands and feet are affected.

The disease is not wholly specific for vitamin B_6 deficiency; it can be produced also in other ways, for example by a lack of unsaturated fatty acids. In various higher animals, convulsions may follow extreme vitamin B_6 deficiency. Sometimes lack of the vitamin causes muscular weakness; it can also cause anemia.

Vitamin B_6 deficiency can be induced by the addition of an antagonistic analogue, desoxypyridoxine. Thus if this compound is injected into hens' eggs, the embryos do not proceed very far in their development. Vitamin B_6 is important for the proper metabolism of the amino acids. It forms part of an enzyme system which promotes the synthesis of amino acids by the addition of carboxyl groups. Acting as part of what is presumably the same enzyme system, the vitamin also favours the removal of carboxyl groups from amino acids.

Another type of enzyme for which vitamin B_6 is an important constituent is transaminase. This enzyme catalyzes the transfer of an amino group from one amino acid to another; an important type of reaction in protein metabolism. Thus there is some important information as to how the vitamin functions with respect to the chemical reactions occurring during the metabolism of proteins and amino acids, but how this is related to the disturbances which occur in skin, nerves, muscle and blood when the vitamin is lacking has yet to be explained. It has been claimed, on the basis of isolation studies of mitochondria, that these granules contain especially large amounts of vitamin B_6. However, it is the nucleoprotein-containing granules of the cytoplasm that are believed to be especially important for protein metabolism.

PANTOTHENIC ACID

Pantothenic acid is found in many diverse types of living materials. Chemically, it consists of β-alanine in a peptid linkage with a dihydroxy acid. (β-alanine differs from ordinary alanine in being β-aminopropionic acid instead of α-aminopropionic acid). The acid is unstable and for this reason is sold as a calcium salt. This salt is a white powder, readily soluble in water but not in alcohol. Many organisms from bacteria and protozoa to higher mammals require pantothenic acid. In its absence these organisms do not grow properly and they may show various ill effects.

Symptoms of pantothenic acid deficiency include dermatitis (inflammation of the skin) in rats, deficient feathering in chicks, graying of hair in rats, degeneration of the nervous system in pigs, hemorrhages in the adrenal gland and general disturbance of function of this gland, etc. The vitamin constitutes part of an enzyme system. Combined with various substances including phosphate, adenylic acid, and a sulfhydryl group, it is coenzyme A, and this in combination with a protein is an enzyme capable of acetylating various organic substances such as choline and sulfanilamide.

The acetylation of choline may be very important physiologically. How the acetylation, or rather the lack of it, can produce all the pathological conditions noted above is not clear, although in the case of the nervous system, any alteration of acetylcholine metabolism would certainly be harmful.

BIOTIN

As long ago as 1898 it was shown that if dogs were fed raw egg white, diarrhea resulted. Following this it was shown that other mammals also suffered from egg white feeding, and that in addition to the diarrhea, skin ailments and nervous symptoms appeared. The phenomenon became known as egg white injury. Cooking the egg white prevented the injury. György proposed the term vitamin H for a factor which prevented egg white injury. Later, it was shown that this factor is identical with biotin, a vitamin known to be important for yeasts and microorganisms.

The raw egg white unites with the biotin and prevents its utilization by mammals or birds. Chemically, biotin is rather complicated. It has in its molecule a urea ring containing one sulfur atom; the ring is combined with a valeric acid group. The presence of sulfur in the molecule is not important for the vitamin activity, for if the sulfur is replaced by oxygen, the vitamin activity is retained. Biotin acts in extremely low concentration, and it is thus one of the most potent of the vitamins. Many organisms do not grow unless biotin is available for them. Usually in higher organisms it is synthesized in sufficient quantity, but biotin deficiency can occur naturally in birds and in insects.

Lack of sufficient biotin has the same effect as egg white injury—diarrhea, disturbances in the skin and nervous system, in rats loss of hair about the eyes. It can also prevent the birth of young in rats.

The protein in egg white which binds biotin is called avidin; it is a basic, carbohydrate-containing protein and it occurs not only in the eggs of birds but also in frog eggs. There is now strong evidence that biotin is important in a process which is known as carbon dioxide fixation. Many bacteria, and higher organisms also, are able to utilize carbon dioxide and form organic compounds from it. This process is interfered with if biotin is lacking.

FOLIC ACID

This vitamin occurs in green leaves; hence the name, which is taken from the Latin word for leaf. Folic acid also occurs in mushrooms and yeast; and in animal tissues, such as liver and kidney. The vitamin has been called by many different names, and it exists in various forms. These have different potencies for different types of organisms.

Chemically, folic acid consists of pteroic acid combined with the common amino acid, glutamic acid. Pteroic acid itself is a combination of para-aminobenzoic acid and a pteridine base. Pteridine bases are pyrimidine derivatives related to the pyrimidine bases thymine and uracil found in the nucleic acids.

The compound shown above is pteroylglutamic acid, sometimes called PGA. Pteroic acid may also combine with 3 glutamic acid groups to form pteroyltriglutarnic acid (P3GA) or with 7 molecules of glutamic acid to form pteroylheptaglutamic acid (P7GA). All of these compounds are found naturally and all have vitamin activity for certain types of living materials. Folic acid is required by some bacteria, by certain ciliate protozoal, by some insects and by various birds and mammals. The bacteria in the rat intestine normally manufacture a sufficient quantity of the vitamin, and treatment with antibiotics such as sulfa drugs is necessary to cause folic acid deficiency.

Folic acid is commonly an essential growth factor, not only for bacteria but for higher organisms as well. Lack of the vitamin in mammals results in various ailments, but especially in a type of anemia characterized by a small number of red blood cells of large size. This type of anemia, called macrocytic anemia, occurs in the tropical disease sprue, and sprue can be cured either by liver extract (which contains folic acid) or by folic acid itself. Folic acid is apparently important both for growth and cell division. When chick embryos are in contact with antagonistic analogues such as 4amino folic acid, their growth is greatly retarded.

Similar results have been obtained in numerous other studies with folic acid antagonists. Thus 4-amino folic acid has a retarding effect on tumor growth. A substance related to folic acid is the so-called "citrovorum factor," which is apparently identical with "folinic acid." This factor favours growth of the bacterium, Leuconostoc citrovorum. The citrovorum factor may be obtained from liver extracts; also it is excreted into the urine of rats or men fed folic acid.

VITAMIN B12

Year by year, the number of B vitamins increases. B_{10} and B_{11} are required by chicks, the former for proper feathering, the latter for growth. From liver it has been possible to isolate a pink crystalline substance which can be used to cure pernicious anemia.

Folic acid is useful in this disease, for it has a hematopoietic effect (that is to say, it favours increased production of blood cells), but it does not correct the nervous symptoms of the disease. These depend on degenerative changes in the nervous system. Vitamin B_{12} in tiny amounts has been thought to correct diseases of both blood and nervous systems, but apparently if the nervous system is badly affected, it can not be cured.

The vitamin has been crystallized. It contains 4-4.5 per cent cobalt and also phosphorus; the molecular weight is approximately 1500. (Estimates vary from 1300 to 1600). In liver extracts there are really two active substances with slightly different absorption spectra. One of these is the definitive B_{12}, the other is B_{12b}. (B_{12a} is another similar substance derived from B_{12} by hydrogenation). Vitamin B_{12} is apparently identical with what has been called "animal protein factor." This was originally obtained from such diverse sources as milk, muscle, liver and cow manure. Although called animal protein factor, it can likewise be obtained from plant sources.

Both animal protein factor and vitamin B_{12} promote growth in some bacteria, in chickens, rats and mice. The effect on growth may be due to the fact that the vitamin seems to increase the utilization of amino acids. At any rate, this has been reported for chicks.

OTHER POSSIBLE VITAMINS OF THE B COMPLEX

B_{13} has been described as a growth factor for rats and for pigs. There is also a description of a vitamin B_{14}, extracted from urine. Over the years, there has been a large and controversial literature about what has been called vitamin P. Substances with vitamin P activity are supposed to maintain proper permeability of the blood capillaries of higher animals and to prevent excessive loss of fluid from them. (The letter P is due to this permeability action). There is now general agreement that substances with P activity are not to be considered as vitamins.

Para-aminobenzoic acid (often called paba). This relatively simple compound is an important growth factor for many bacteria. Thus sulfanilamide is an antagonistic analogue of para-aminobenzoic acid, and its antibacterial action is due to this similarity in structure. The discovery of this relationship in 1940 was indeed a milestone in our understanding of the mechanism of drug action and has served as a model and guide for many subsequent researches.

Choline

This substance, the basic constituent of lecithin, is believed by many to be a vitamin. Rats deprived of choline develop fatty livers and other types of hepatic injury. They may also show degenerative changes in the kidney. Choline deficiency is also a factor in the development of a leg deficiency (perosis) in chicks and turkeys.

It should be remembered that choline is a constituent not only of lecithin, but also of the important humoural substance, acetylcholine. For a discussion of the significance of acetylcholine, *Inositol*. Inositol is a derivative of benzene.

It can be prepared from hexahydroxy-benzene by reduction, in the course of which 6 hydrogen atoms are added. Thus inositol is hexahydroxy-hexahydro-benzene. It is an optically active substance found in yeast, muscle, and in various other types of animal and plant tissue. Inositol is important for the growth of yeasts, and is indispensable for some types of yeast.

It also promotes the growth of some fungi. In the presence of inositol, chicks grow more rapidly than when it is absent. However, rats may be bred for three generations without inositol, and György was led to conclude that "the facts presented do not, at the present time, warrant the identification of inositol as a primary and really essential vitamin; they rather favour the assumption that it is a supporting and at least not always specific vitamin." *α-Lipoic Acid*. This substance, also called protogen, is probably to be included among the vitamins of the B group, for it is water soluble.

It is important for the growth of some bacteria and for the protozoan Tetrahymena, and it is involved in the oxidation of the important keto acid, pyruvic acid (for this reason it was formerly called pyruvate oxidase factor). Many types of living material contain lipoic acid. It is now known to be a derivative of octanoic (caprylic) acid and it contains sulfur in the form of an S-S group.

There are doubtless other vitamins of the B group still to be discovered. Fraenkel found that the larva of the beetle Tenebrio molitor (the so-called mealworm) requires some substance present in yeast or liver but not one of the known B vitamins. This he called B_T. More recently he has identified it with carnitine, a dipeptid known originally in muscle.

Also the flour beetle requires at least two other unidentified members of the B complex and neither of these is B_T. Goetsch claims that there is a vitamin present in certain insects, especially termites, a vitamin which has pronounced effects on growth and differentiation. This he calls the vitamin T complex; it is supposed to affect not only insects but various other types of organisms. Throughout the course of vitamin study, there have indeed been many extravagant claims and many false leads. The vast literature is often contradictory and confused. But in spite of this, in the course of time the truth has gradually emerged and the main outlines of the picture are now clear.

All sorts of organisms, from bacteria to man, require certain specific types of organic molecules. These can be synthesized by green plants and by many lower organisms. Higher forms have frequently lost the power for such synthesis, and unless, like the ruminants, they provide themselves with huge storehouses for bacteria, they must depend for their existence on obtaining the vitamins with their food.

From the knowledge of the nature of the vitamins, we can begin to develop a chemical anatomy of living systems. The vitamins represent indispensable organic molecules and apparently the same molecules are almost universally needed. Moreover, we are beginning to discover why the vitamins are needed and what work they do in the living cell. Much of our information is still fragmentary.

The vitamin requirements for the lower vertebrates and for the hosts of invertebrate animals have scarcely been investigated at all. Moreover, in some instances we are completely at a loss to account for the fundamental necessity for a given vitamin.

Such superficial disturbances as a reddening of the eye or a scaliness of the skin give no direct indication of the part a vitamin may play in protoplasmic activity.

Chapter 6

Absorption

In those organisms in which food is taken into an alimentary canal and then digested, the products of such digestion must pass through the walls of the alimentary canal in order to find their way to the tissues and cells. This introduces the question of intestinal absorption. The problem of the mechanics of intestinal absorption has had a fascination for many physiologists. In the study of such absorption there is a possibility of applying physicochemical knowledge.

Thus, it is obvious that digestive changes, which convert indiffusible materials such as proteins and starches into diffusible compounds of much smaller molecular dimensions, make possible the passage of these now diffusible compounds through the membranous lining of the intestine.

But although the laws of diffusion, and the laws of osmosis, are important factors in explaining intestinal absorption, there remain facts which cannot be explained on any simple physicochemical basis. To arrive at a complete understanding, either one must assume strange and relatively unknown physical processes, or one can, if one please, more or less beg the question by assuming vital or undetermined biological factors. In higher animals the lumen of the intestine is lined with finger-like projections or villi. These have muscle fibers in their walls, and they appear to be in a constant state of motion, now shortening, now lengthening.

Each villus seems to contract independently, and in dog intestines the rhythm of contraction is at the rate of about 6 per minute. These contractions were first clearly noted by Hambleton in 1914. Since then they have been carefully studied by Verzár, and by various other workers. Contractions of the villi seem to have some significance for intestinal absorption. If the movement of villi is stimulated, intestinal absorption increases. Thus if HCI is added, there is an increase in rate of movement and an increase in the rate of absorption of glucose. (The effect is believed to be due to a hormone produced in the intestinal wall by the acid).

Condiments such as pepper, paprika, onion and garlic also cause an increase in the rate of movement of villi. Conversely, if cocaine is added, the movement of the villi may be prevented and then absorption of glucose drops 12-30 per cent. It thus seems clear that the villi play an important part in the absorptive activity of the mammalian gut. As an argument against the importance of the villi it has been claimed that in the case of animals without intestinal villi the same type of absorption occurs.

This claim, based on widely quoted experiments of Cohnheim on the isolated intestine of sea cucumbers, is apparently invalid. In the study of intestinal absorption, food materials or chemical substances of one sort or another are inserted into a loop of the intestine, and then after an interval of time the contents of the loop are examined in order to determine how much water and dissolved substance have escaped from the interior of the loop. Either the animal is anesthetized and operated on and the desired portion of the intestine tied off, or an isolated portion of the intestine is permanently sewed to the abdominal wall (*Vella fistula*).

Both methods have advantages and disadvantages, but in both of them there is some tendency for the intestinal loop to be under abnormal conditions which may easily affect movement of the villi, etc. A third method is that of Cori. In this method a small amount of the substance to be investigated is fed by stomach tube to a small animal, e.g., a rat. After a given time the animal is killed and the amount of substance remaining in the intestine is determined quantitatively.

The difference between the amount fed and the amount recovered from the whole intestinal tract is the amount of substance absorbed. Although with Cori's technic it is possible to study absorption under normal conditions, there may be variation in the length of time the introduced substances remain in the stomach before passing to the intestine, and the time factor for absorption is therefore uncertain. In the human, it is possible to insert tubes not only into the stomach but into the intestine as well. With the aid of such tubes samples of stomach or intestinal contents can be analysed at intervals following the introduction of sugars or other food substances.

In this way intestinal (and stomach) absorption can be studied in a relatively normal subject, but it must be remembered that the movement of villi which is under nervous control may be influenced to some extent by any irritation produced by the insertion of the tubes. There have been many studies of nutrition based on the use of isotopes, and these nutritional studies often give information concerning intestinal absorption. Deuterium (heavy hydrogen) has been widely used and also radioactive isotopes of phosphorus, sodium, etc.

In these studies, which will be referred to again later in the chapter, the isotopes or compounds containing them are introduced into the intestine and their presence is then recognized in blood or tissues. Sometimes the procedure may be rather simple. Thus a man can be fed radioactive sodium and following suitable intervals of time the presence of radioactive sodium in his hands can be determined with the aid of a Geiger counter. Quantitative study of intestinal absorption has been confined almost entirely to the vertebrates, and most of the work has been done on mammals. This is unfortunate, for comparative data on absorption rates in different animals might be of considerable interest.

According to Jordan and Begemann the intestine of the snail (Helix pomatia) is more permeable to disaccharides than is the intestine of the frog. In some invertebrates (sponges, coelenterates, flatworms, echinoderms) intake of food from the body cavity or intestine is largely controlled by phagocytosis.

This is not true for insects. As an example of an absorption experiment, the following table is taken from Höber. It represents an experiment of Heidenhain's on an intestinal loop of a dog. The ends of the loop were tied off, but the loop itself was provided with blood through the mesentery. The record shown in the accompanying table indicates that the membrane lining the intestine does not behave as a semipermeable membrane, and that the laws of osmosis are obviously not followed.

Thus, when solutions of sodium chloride more concentrated than the blood are introduced into the loop, they lose water and, moreover, sodium chloride passes out of the loop. Nevertheless, it is obvious that osmotic forces are not without effect, for the stronger solutions of sodium chloride lose much less water than do the weaker solutions.

Intro duced NaCl	*Into Loop Removed NaCl,*	*Time For Gm.*	*From Loop cc. Percentage Min.*	*Total NaCl.*	*Absorption. cc. Percentage NaCl,*	*Total Gm.*
120	0.3	0.36	15	18	0.60	0.108
120	0.5	0.6	15	35	0.66	0.23
117	1.0	1.17	15	75	0.90	0.67
120	1.46	1.75	15	109	1.20	1.31

The rate of absorption of various types of dissolved substances from the lumen of the intestine has been studied by many authors. Some of the papers are referred to in the reviews of Goldschmidt, Magee and Verzár, and the monograph of Verzár and McDougall. Salts such as magnesium sulfate and sodium sulfate pass through the intestinal wall much less readily than sodium chloride; they cause a flow of water into the intestine and thus produce a laxative action. Höber and Höber have studied the absorption of various organic substances in solution. For substances of no physiological importance, rate of absorption decreases with rising molecular volume.

However, substances of distinct lipid solubility are absorbed at a higher rate than those not soluble in lipids. There is thus a similarity to cellular permeability. Höber and Höber found that amino acids were absorbed at a higher rate than non-physiological substances of the same molecular volume.

Numerous authors have compared the rate of absorption of various simple sugars. A similar series has been described for the intestines of the pigeon and the frog, but in the cat intestines, xylose is absorbed as readily as glucose. This paper has many references to the literature. In general, the sugars most readily absorbed are those most needed in the body.

According to Verzár and McDougall: "In the absorption of certain substances there is a special selective advantage which seems to be completely contrary to physical laws. These 'favoured' substances are ones which are especially used in the metabolism of the body. Such a specific selectivity is the result of active life processes in the mucosal cells.

Thus from a mixture of hexoses and pentoses, glucose and galactose are selectively and quickly absorbed (three to four times as quickly as the others from blood-isotonic solutions)." The phosphorylation theory proposes an explanation of the selective absorption of sugars.

It was suggested originally by Wilbrandt and Laszt that the reason hexose sugars were absorbed more readily than pentoses such as xylose (in spite of their greater molecular volume) was the fact that hexoses were able to combine with phosphate.

In the cells of the intestinal wall such "phosphorylation" was believed to occur and to be responsible for the more rapid absorption of glucose. The phosphorylated sugars must be dephosphorylated again as the sugars pass into the blood from the intestine. The evidence in favour of the theory has been many-sided.

Wilbrandt and Laszt showed that the presence of phosphates in the intestine favours the absorption of sugars and that iodoacetic acid which prevents phosphorylation also prevents rapid absorption of the hexoses. Similarly, according to Lundsgaard, the alkaloid phlorhizin prevents phosphorylation and also the selective absorption of hexoses. Removal of the adrenal gland is also believed to interfere with phosphorylation and with the absorption of sugars. The literature on phosphorylation is voluminous and not wholly concordant.

The older papers are referred to in Verzár and McDougall's monograph, already cited. Objection has been raised that iodoacetic acid is a serious poison and does much more than merely prevent phosphorylation. Moreover, there is some question as to whether phlorhizin is an agent which specifically inhibits phosphorylation in vivo. On the whole the evidence indicates that the phosphorylation reaction hastens the absorption of sugar.

Kjerulf-Jensen and Lundsgaard attempted to study the rate of dephosphorylation in the intestine in order to discover whether phosphorylation-dephosphorylation reactions could be rapid enough to explain selective absorption. They injected glucose into the intestines of rats, killed the animals 15-20 minutes later with strong cyanide solution, and then inasmuch as the cyanide prevented phosphorylation, they were able to follow the course of dephosphorylation.

Phosphate esters present in the intestines of sugar-fed rats were able to release inorganic phosphate rapidly. Within one or two minutes the reaction went to completion. Kjerulf-Jensen and Lundsgaard conclude that the phosphorylation-dephosphorylation reactions in the rat intestine are rapid enough to account for the phenomenon of selective absorption of sugars, but they do not regard their results as of necessity proving that phosphorylation plays the essential role in the absorption of hexose sugars, for it is possible that the phosphorylation reaction may merely provide the source of energy.

In an earlier paper, Lundsgaard was not able to obtain much evidence for phosphorylation in the intestine of the cat. This may be correlated with the fact already noted that the cat intestine shows no selective absorption of hexose sugars. Ideas on fat absorption have changed with the years.

Before 1900 it was commonly held that some of the fat taken into the body of an animal was absorbed only after it had been hydrolyzed into fatty acid and glycerol, but that in addition some fat could be absorbed as finely emulsified particles. In 1900, Pflüger argued strongly that *all* the fat had to be hydrolyzed in order to be absorbed and that no emulsified fat could be taken into the wall of the intestine. This then became the accepted belief and found its way into all the textbooks. In 1940 Frazer was impressed with the fact that fat could be absorbed in the form of finely emulsified droplets.

An interesting argument in favour of the absorption of emulsified droplets of oil is the fact that paraffin oil is not ordinarily absorbed, but if it is emulsified by the addition of small amounts of oleic acid and cholesterol, then it is taken into the intestinal wall. As Frazer found more and more evidence that emulsified fat could be absorbed, he came more and more to doubt whether any large amount of fat was absorbed in the form of fatty acid and glycerol.

Thus he points out that higher fatty acids are insoluble in water. Of course the salts of the fatty acids (soaps) are highly soluble, but Frazer states that at the slightly acid *p*H of the small intestine of higher animals, the amount of soap that can be present is negligible. (Actually, even at *p*H 6, there is an appreciable amount of fatty acid present in the form of soap). The emulsification of fats is favoured by bile and by the lecithin that is typically present in bile. Frazer believes that the tiny droplets of fat enter the cells lining the intestinal epithelium through pores present in the outer boundary of these cells; these pores lead to canals, and the canals lead to the naked protoplasm.

As a matter of fact, it has been observed both for plant cells and for simple animal cells such as sea urchin eggs that if an oil droplet is placed in contact with the cell membrane the oil will rapidly enter the cell interior. Frazer may go too far in claiming that practically all fat is absorbed in emulsified form. No doubt some is absorbed in this way, but probably a very appreciable amount is also absorbed as fatty acid and glycerol. This is strongly indicated by an isotope study of phospholipid absorption made with the aid of radioactive phosphorus.

In this study it was clearly demonstrated that some of the lipid was absorbed without hydrolysis having taken place, but that some of the lipid was absorbed after hydrolysis. A similar result was obtained in an isotope study of the absorption of triglyceride fat. Phosphorylation is also believed to be important for the absorption of fats. The idea is originally due to Sinclair.

He thought that fatty acids entering the intestinal wall combined with phosphate (and a base) to form phospholipids and were then split off again to form neutral fat. Much of the evidence for the importance of phosphorylation in fat absorption is similar to that advanced to show the importance of phosphorylation for carbohydrate absorption; it is reviewed in Verzár and McDougall's monograph, also by Bloor.

In a study with radioactive phosphorus, Zilversmit, Chaikoff and Entenman concluded that the formation of phospholipid was not a necessary step in the absorption of fat. In other words, fat can be absorbed even if no phosphorylation occurs. Concerning the mechanism of absorption and the forces responsible for passage of water and dissolved materials through the intestinal epithelium, there has been wide diversity of opinion. Goldschmidt cites 15 different types of theoretical explanation. After summarizing the evidence, Magee in his review states: "it will be seen that the majority of the experimental results cannot be explained by known physical laws.

The conclusion may therefore be drawn that the phenomena of absorption are due partly to some special property of the epithelial cell." One difficulty that all theories have to face is the fact that active osmotic work occurs during intestinal absorption. The energy for this work presumably comes from the protoplasmic activity of the intestinal cells. But what sort of activity is involved and how is the energy applied?

An attempt to find some sort of answer to this problem was made by Visscher. He and his collaborators have carried out thorough and extensive investigations on the problem of salt absorption, and they have shown very clearly that other than osmotic forces must be involved.

Ingraham, Peters and Visscher adopted what they called the "fluid circuit" theory. This theory maintained that an ion such as chloride passed in through one region of the intestinal membrane and that water passed out through another region. The driving force of such an arrangement is not clear. In their discussion of theory, Peters and Visscher stated that there were five general types of theory (the fluid circuit theory constituting the fifth).

They easily disposed of the first three types of explanation, and the fourth which postulates some type of compound formation, they attacked on the ground that no substance was known which would combine with NaCl.

However, this is hardly a fair criticism, for the proteins of the cell certainly can combine with cations such as sodium, and there is also a possibility that they may combine with chloride. As a matter of fact Hanke and Donovan claimed that 35 per cent of the chloride of the intestinal wall was bound. In view of the fact that in living systems practically all the energy produced in the course of vital activity is derived from chemical reactions, it seems logical to assume that the energy required to transport dissolved substances into the intestinal wall against osmotic forces is essentially chemical.

Franck and Mayer have considered the problem from a thermodynamic standpoint and they have postulated an osmotic diffusion pump which would obtain energy from chemical reactions and would be able to transfer either solvent or solute across membranes. In their discussion of the types of reactions which might conceivably be involved, Franck and Mayer consider both phosphorylation reactions and the combinations of ions with substances like resins.

Of course there are no resins in ordinary protoplasm; but, as already pointed out, ions can and do combine with proteins, and with proteins as with resins exchange of ions is possible. The absorption of calcium presents a special problem and there is a large literature on it. This ion, so important in vital processes, is absorbed only with difficulty from the intestinal tract. Diseases such as rickets are due to faulty absorption of calcium. The fact that vitamin D favours calcium absorption can perhaps be explained in terms of the effect of the vitamin on the combining power of calcium with the proteins or lipids of the cell.

This aspect of the subject needs further investigation. In the physiology of mammals and higher animals generally, there is not only the problem of absorption into the blood or lymph from the intestine, but also the problem of how the tissues absorb dissolved substances from blood or lymph.

This latter question has received relatively little attention. Cori and Goltz studied the absorption of sugar by the liver and muscles from the blood. The absorption rate was rapid and approximately the same for both types of tissue.

The absorption of food materials by the cells of animals is to some extent paralleled by the absorption of food by the plant. In plants, generally, there is no necessity for digestion of solid food, and only in insectivorous plants does this occur. But in so far as the intake of food into the cells themselves is concerned, both in plants and animals there is evidence of a selective action on the part of the living cells. In plants also, there may be an active transport of dissolved substances against a diffusion gradient.

This is indeed a characteristic of many, perhaps all, types of living cells and is not confined to the cells which are concerned with the uptake of food in animals and plants.

Chapter 7

Fluid Transport in Body

We may begin with the exchange of gases, which in our own case we call breathing. In simple animals like jellyfish and many worms there are no special organs for this purpose. Dissolved oxygen soaks through their skins from the water in which they live, and dissolved carbon dioxide soaks out. In somewhat more complicated creatures like the earth-worm there is a special fluid, the blood, one of whose functions is to carry oxygen from the skin to the internal organs, and carbon dioxide back.

In the most advanced water-dwellers we find special tufts of thin skin, the gills, into which blood is pumped by a heart or hearts, so as to expose it to water with which it may exchange gases. Gills may be naked, as in the young tadpole, but usually, as they are very delicate, they are protected by lids or pouches, as in fish or lobsters, where they lie behind the head on each side. In air-breathing animals there are two quite different kinds of breathing organ. In insects and spiders, air is carried to every part of the body by tiny branching tubes called tracheae, which open by numerous pores, mainly on the sides of the abdomen.

In air-breathing vertebrates and molluscs the blood is exposed in the lungs to air which is continually renewed by the act of breathing. Some animals combine several methods. Thus a frog can use its skin alone for breathing so long as it does not need much oxygen, but it normally employs its lungs as well. The human lungs are elastic organs of a spongy texture, consisting of millions of very small cavities which open into stiff tubes called bronchi. These in turn open into the windpipe in the front of the neck. Since air cannot enter the space between the lung and the chest-wall, the lungs expand when the chest is expanded.

This can happen in two different ways. In the first place the diaphragm, a sheet of muscle separating the chest and belly, and bulging upwards into the former, may contract and force the contents of the belly downwards and outwards.

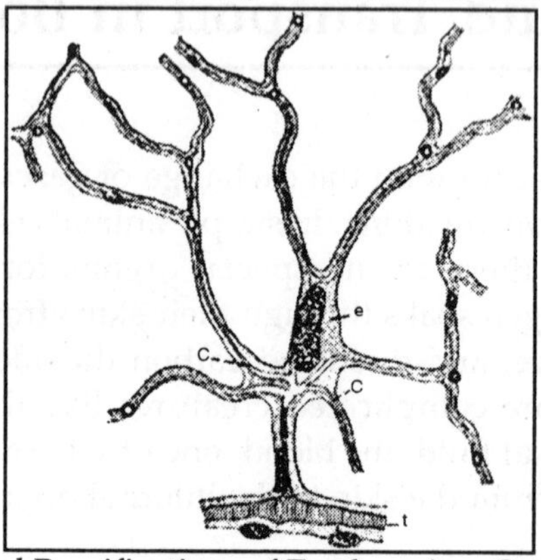

Fig. Final Ramifications of Tracheae in a Caterpillar.

This pulls the bases of the lungs down, and draws air into them. Or the muscles which lie between the ribs may contract so as to bring the ribs, which normally slope downwards, into a more horizontal position. The breast-bone is thus pushed away from the vertebral column, while at the same time the diameter of the chest from side to side is increased, and the lungs expand to fill the extra volume.

Muscles acting in the opposite direction force the air out again if necessary; generally the elasticity of the lungs is sufficient. Thus the air to which the blood in the lungs is exposed is constantly being changed.

The inspired air contains 21 per cent. of O_2 and only 0.03 per cent. of CO_2, whilst the expired air (after being dried) contains about 17 per cent. of O_2 and 3 per cent. of CO_2. To understand how this change occurs we must study the circulation of the blood. The blood flows to and from the heart, as we have seen, through a closed system of tubes known as arteries, veins, and capillaries. It leaves the heart by the arteries, which are comparatively thick-walled, and seldom near the surface, and returns by the veins, which have thinner walls, and often lie just below the skin.

It passes from the arteries to the veins by the capillaries; which are too small to be seen with the naked eye, but are found in almost every tissue. The most easily felt arteries in man are those of the wrist and temple, the most easily seen veins those on the back of the hand and foot. Finally, the heart, which pumps the blood round, is a hollow muscle with four chambers, lying between the lungs, and about as large as the two fists together. The blood from all parts of the body except the lungs enters the heart from behind, near the right-hand top corner, and flows into a chamber called the right auricle.

This contracts rhythmically (the average rate in a grown man at rest is about seventy times a minute) and forces its contents into a thicker-walled chamber lying below it, the right ventricle. As soon as this is full it contracts in its turn. The blood is prevented from returning into the auricle by a valve, called the tricuspid. It therefore finds its way out through the pulmonary artery, which leads it to the lungs. The semilunar valves at the base of the artery prevent it from flowing back into the ventricle when this relaxes at the end of its stroke. In the lungs it has to pass through capillaries in the walls of the air-sacs, so that it is only separated from the air by a very fine membrane, and can easily lose its carbon dioxide and take up oxygen.

It returns from the lungs by the pulmonary vein, this time to the left auricle, which contracts at the same time as the right auricle, and fills the left ventricle. The left ventricle, which is assisted by valves not unlike those on the right side of the heart, forces the blood into the largest artery, an elastic tube called the aorta, from which it is distributed all over the body by the other arteries.

One way of forcing it round through the narrow capillaries would be to have a cistern at the top of the head from which it flowed down again, but this arrangement would clearly not work when its owner lay down. In reality, a fairly steady pressure is kept up by the fact that the walls of the aorta and the arteries are always stretched, and continue to squeeze the blood along between the strokes of the ventricles.

The pressure in man is about that of a column of blood 1.7 metres high, i.e. equal to the head of blood which would be obtained from a cistern 1.7 metres above the heart. The heart of a grown man at rest delivers about 8 litres per minute, or just over 100 cubic centimetres per beat.

During exercise the heart may deliver three or four times as much per minute, mostly as the result of an increased number of beats per minute. The beat of the heart which is felt below the fifth rib on the left side is due to the left ventricle striking the wall of the chest each time it contracts and stiffens. The blood is squirted along the arteries at a rate which may be as high as 50 centimetres a second, and each fresh wave causes a pulse in the artery.

If an artery is cut, the blood comes in a series of spurts from the side nearest the heart, and can be stopped by pressure on the heart side of the cut. It is easy (and safe) to stop the pulse in the wrist by pressing on the same artery higher up, either inside the elbow or inside the upper arm just below the armpit, where it can be felt. The blood flows through the capillaries very slowly, at about half a millimetre per second. Their average length is about a millimetre, while their diameter may be less than 0.01 millimetre. Their very thin walls allow water, gases, and dissolved substances to be exchanged between the blood and the tissues with great ease. From the capillaries the blood oozes gently into the veins.

They have no pulse, and a comparatively small pressure and rate of flow. A cut vein bleeds steadily and the flow can be stopped by compressing the side away from the heart. The flow of blood in the veins is assisted by the presence of valves in them, which only allow it to move towards the heart. If the finger-tip be run along one of the veins of the forearm away from the heart, the vein will dilate on the heart side of each valve.

When therefore the contraction of various muscles squeezes the veins, the blood can only flow towards the heart. If a man stands quite still the blood tends to accumulate in the veins of his legs, and he is liable to faint from failure of the supply to his brain. When he starts walking the blood is at once squeezed out of these veins. The only veins in man which do not lead straight back to the heart are those from the digestive canal and some other abdominal organs, which pass into the liver.

Here the food absorbed by the blood from the gut is dealt with. As the liver needs oxygen, it has also a supply of fresh arterial blood. In most other animals the circulation is somewhat different. Thus in a fish the blood only goes once through the heart in a complete journey, instead of twice, and it all passes through the gills before going on to the tissues. In the frog, in keeping with its two ways of breathing, we have a condition almost half-way between that found in fish and men. The blood entering the right side of the heart by the main vein flows into *R.A.*, the right auricle; thence, through a valve, into *R.V.*, the right ventricle, which pumps it, through a valve, down *P.A.*, the pulmonary artery, to the lungs, *Lg*.

The blood here flows in capillaries, and is then gathered up into the pulmonary vein, *P.V.*, and taken to the left side of the heart, entering the left auricle, *L.A.*, then (through a valve) the left ventricle, *L.V.*, and by this being pumped out (again through a valve) into the aorta, *Ao.* Some of the blood (A^1) goes to the head and anterior extremities, whence it is returned by the anterior venae cavae (V^1), into which discharges the main trunk of the lymphatic system, the thoracic duct (*Th.D.*). The rest (A^2) goes to the trunk and hind limbs.

That which supplies the digestive tube (*Al.*) passes to the liver (*Lr.*) by the hepatic portal vein (*V.P.*); the liver also receives arterial blood direct by *H.A.*, the hepatic artery. The hepatic vein, *H.V.*, joins the veins from the other posterior regions, V^2, and flows to the heart in the inferior vena cava, *V.C.I. Lct.* denotes the lacteal lymphatic vessels from the intestinal wall, *Ly.* ordinary lymphatics. We must now consider how the blood acts as a carrier of oxygen and carbon dioxide.

Water at body-temperature will only take up one two-hundredth of its volume of oxygen from air; if the blood were no better than this we should need a heart working forty times as fast as the one we have. Actually the oxygen is carried round in loose chemical combination with a body called haemoglobin. This is a protein, but contains iron as well as the usual carbon, hydrogen, oxygen, nitrogen, and sulphur. It is of a purple colour, but becomes red on combining with oxygen, which it readily takes up from the air.

Thus the blood in the veins is purplish, but if exposed to air either in the lungs or by opening a vein, it at once becomes red. The haemoglobin which it contains enables blood to hold 18 volumes of oxygen per cent., which is 6/7; of the amount contained in the same volume of air, and about 40 times what the blood could carry without haemoglobin. Since blood carries dissolved solids as well as gases, a single set of capillaries supplies the tissues with all that they need for activity, growth, and repair, besides removing most of the waste products.

The carbon dioxide produced in the tissues is mainly carried, not in solution as such (CO_2), but in combination as sodium bicarbonate ($NaHCO_3$). If we look at a drop of blood under a microscope we see that it consists of a clear fluid full of little reddish-yellow biscuit-shaped bodies about 0.007 millimetre in diameter. They can just be seen with a powerful hand-lens. A cubic millimetre of blood contains about five million, and as a man has about 4 litres of blood the total number in his body is about 20 million million, or far more than the number of men who have lived since history began, or probably at all. Their total surface is about 1,500 times the surface of the body.

All the haemoglobin of the blood is contained in them, and they therefore carry round the oxygen from the tissues. Their huge surface area renders this exchange easy. They are cells which have lost their nucleus and are therefore not fully alive, but act as passive carriers of oxygen. They are always being produced in large numbers in the bone-marrow, and destroyed in the spleen and liver when worn out.

For every 500 or so of these red corpuscles there is one white corpuscle. They are true cells with nuclei, and some of them are capable of active movements. There are at least six different kinds with different functions, mostly of a protective character.

Thus one kind eat up parasites found in the blood, another kind burrow through the walls of capillaries in inflamed areas, and remove dead or injured tissue and disease-germs, often forming collections of 'matter' or pus. Others produce substances which kill disease-germs, and so on. The blood also contains non-cellular bodies called platelets, which are smaller than the corpuscles and are concerned in clotting and in immunity; there may also be tiny drops of oil after a fatty meal. Besides the blood-vessels, all vertebrates possess another system of vessels which open into the veins, but contain clear fluid called lymph.

The lymphatic vessels slowly drain away fluid from the spaces between the cells. On the course of the vessels the fluid passes through small lymph-nodes or glands, which may be felt under the skin in the neck, armpit, or groin, before entering the blood. The lymph contains white corpuscles which are largely produced in these glands, and abnormal bodies from the tissues are dealt with in them. Thus if the arm is inflamed, the lymph-nodes of the armpit are generally enlarged, as they are busy destroying poisons or bacteria from the inflamed tissue.

The lymphatics also play a part in digestion, which we shall study later. We must now see how the cells obtain the food-stuffs which they require. In simple animals such as polyps every cell is so close to the digestive cavity that it can obtain food thence directly. In most higher animals, however, the cells which line the alimentary canal and its glands are highly specialized for the purpose of breaking up the food-stuffs into soluble forms and passing them rapidly into the blood or other body-fluids.

This process is called digestion. A few internal parasites like the tapeworm have no gut, but absorb their food through their skin, relying on their host to break it up for them. The course of digestion in man is as follows. The food is chewed in the mouth, where it is also moistened by saliva, a fluid secreted mainly by three neighbouring pairs of glands. Two pairs of these lie between the tongue and the lower jaw, the third pair (whose inflammation causes mumps) lie below the external ear, mainly inside the lower jaw-bone.

Besides moistening the food, saliva contains an enzyme, ptyalin, which breaks up starch into an easily soluble sugar called maltose. Enzymes play a much larger part in digestion than do mechanical processes. Each digestive enzyme is a definite substance with the property of bringing about, or enormously speeding up, a particular chemical reaction.

Pepsin from the stomach will split up half a million times its weight of protein, but will not alter starch or fat. So delicately is an enzyme adjusted to the substrate on which it acts, that as most food molecules are asymmetrical (as we know from their rotation of the plane of polarized light and their asymmetrical crystals) so are the enzymes that act on them.

Enzymes have been compared to keys which open certain locks only. But they are like 'Yale' rather than ordinary keys. For we can make in the laboratory a sugar or peptide (part of a protein molecule) which only differs from the natural variety in that its molecules are related to the natural molecules as a left hand to a right or an object to its image in a mirror. Enzymes will not act on these artificial substances, though they digest their mirror-images which occur in nature. So Alice, who went through the looking-glass in the story, could not have digested the looking-glass proteins and carbohydrates.

She would have had to get her energy from fat and alcohol, whose molecules are symmetrical, and would finally have died of protein starvation. Enzymes are not alive, and can still work when removed from the body; but they are generally destroyed by boiling. The action of ptyalin can easily be shown. A solution of boiled starch gives a blue colour with iodine.

If saliva is allowed to act on it for a few minutes at body-temperature or a few hours in the cold, the starch is broken down to sugar and loses this property. If the saliva is first boiled no change occurs. When the food is swallowed, it is passed into the gullet or oesophagus. To get there it must pass over the mouth of the wind-pipe, and when any falls down this we choke.

To prevent this, the breathing is stopped while we swallow, and the cartilages at the top of the wind-pipe which are concerned in voice-production are brought together so as to close the top of the larynx. It is further protected by a cartilaginous lid at the back of the tongue called the epiglottis, which is pressed backwards over it by the food during swallowing. The movements of the laryngeal cartilages can easily be felt. As the food leaves the mouth it passes out of the control of consciousness and will. The movements of our digestive canal, except at the two ends, are carried out by smooth or involuntary muscle and are controlled by a special part of the nervous system over which the mind does not preside.

The food or drink on leaving the mouth is seized by the smooth muscle of the oesophagus, which contracts behind it and relaxes in front, thus passing it rapidly down into the stomach. Owing to this gripping action one can eat or drink while standing on one's head.

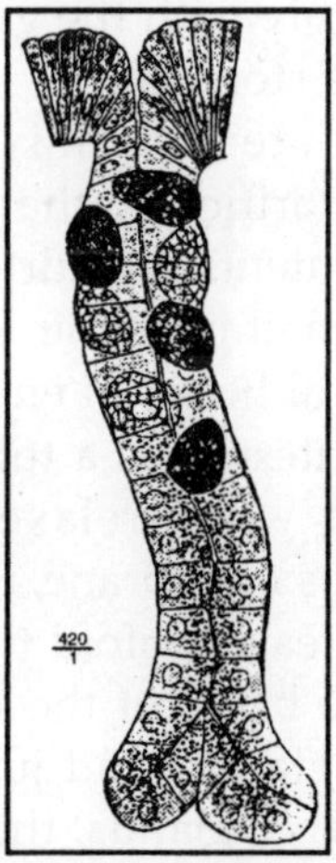

Fig. A Simple Gland From The Stomach of a Bat.

The narrow duct opens above into the cavity of the stomach, among columnar cells. The main part of the gland is a simple tube, formed of a single layer of cells; these secrete pepsin. A few darker-stained cells are seen near the outer side of the tube; these probably secrete hydrochloric acid.

The human stomach is a bag of smooth muscle lined with a membrane consisting mainly of microscopic glands. When expanded it will generally hold about 2 litres. The gastric juice is a clear fluid containing about ½ per cent. free HCl, and several enzymes, of which the most important is pepsin. In presence of acid (though not in a neutral or alkaline fluid), pepsin causes proteins to break up into bodies called peptones and proteoses, which have very much smaller molecules than the proteins from which they are derived, and are more soluble.

Some carbohydrates are attacked by the hydrochloric acid. Thus each molecule of cane-sugar is split into one of glucose and one of fructose, and inulin, a starch-like body found in many plants, is broken up into fructose. Fats are less affected in the stomach. A meal remains in the stomach for a time—generally between one and four hours—which depends on the nature and quantity of the food taken. All this time the muscular walls of the organ are contracting in such a way as to mix the food thoroughly with the gastric juice. Hardly any absorption occurs in the stomach.

When its contents are fully mixed the muscular ring surrounding the lower orifice of the stomach, the pylorus, opens, and a jet of its contents is squirted into the duodenum, the top portion of the small intestine. After a heavy meal the stomach may take several hours in emptying itself.

The human small intestine is a tube about 6 metres long and 3 cm. in diameter when relaxed. It is lined by fine projections of the mucous membrane, the villi, which give it a velvety texture, and increase tenfold the surface available for absorption. Between the bases of the villi are the mouths of numerous microscopic glands, and just beyond the pylorus open the ducts of two large glands, the liver and pancreas.

The liver, among its other functions, secretes bile. This contains substances which have the property, possessed in a less degree by soap, of lowering the surface-tension of water in which they are dissolved. Drops of oil or melted fat tend to join up so as to have as small a surface as possible. This tendency of the surface to shrink is prevented by the bile in the intestine, so when fat-drops are broken up there they do not coalesce again, but form a milky emulsion. Hence the fat-splitting enzyme made in the pancreas can act on a vastly greater surface of fat than would otherwise be available. A man with jaundice from a blocked bile-duct can digest milk, whose fat is already broken up, but not suet or butter, which form big drops.

The bile also contains pigments formed from the haemoglobin of worn-out red corpuscles. These are excreted in the faeces and give them their yellow colour. In jaundice the bile-duct is blocked, and accordingly the skin becomes yellow and the faeces white. Bile is stored in the gall-bladder till required.

The pancreas secretes a juice containing a number of enzymes. One of these, like ptyalin, breaks down starch into maltose. Others break down maltose to glucose, and lactose (milk-sugar) into glucose and galactose, a very similar substance. Another breaks down fat into glycerol and fatty acids, which partly combine with alkali to form soap.

In every case the molecules formed will pass more easily through a membrane than those of the food. If the juice contained a protein-splitting enzyme it would probably digest the pancreas itself and its duct. However, it contains a substance which, on mixing with the secretions of the intestinal glands, yields trypsin, an enzyme which attacks proteins and the products of peptic digestion, breaking them down into amino-acids. These enzymes will not act in an acid fluid, so the bile and pancreatic and intestinal juices contain enough sodium bicarbonate to neutralize the acid of the gastric juice. The secretion of the intestinal glands, besides the substance which helps to form trypsin, contains an enzyme which will break down peptones, though not whole protein molecules, into amino-acids.

There are also enzymes which break down milk-sugar and cane-sugar to sugars containing only six carbon atoms. So far as we know, no animal more complicated than a snail produces an enzyme which will break up cellulose. But grass consists very largely of cellulose, and to digest it hoofed animals and other plant-eaters employ bacteria which grow in their digestive canal. In cud-chewers like the cow and sheep these bacteria live in special compartments of the stomach; in the horse in the large intestine, which may have a capacity of 200 litres.

The bacteria can get very little oxygen, so they cannot oxidize the cellulose, but they turn a good deal of it into methane, which is wasted, and leave much undigested. The rest, however, is broken up into small molecules which the animal can absorb. In man, cellulose is not digested, but it is useful in giving bulk to the faeces and preventing constipation, which easily occurs when the food leaves no indigestible residue. The food is thus broken up into easily soluble constituents, and is ready to be absorbed. This is done by the epithelium of the small intestine. The passage through is not a mere filtration. For example, the blood contains one part of sugar per thousand, and if sugar merely filtered across from gut to blood, it could never quite disappear from the gut as it in fact does.

Actually, during the absorption of food, the absorptive cells perform work, for which they need an extra supply of oxygen. The fats are, in part at least, put together again from glycerine and soap, and passed as a fine milky emulsion into the lacteals, as the lymphatics of the small gut are called. These join together to form the thoracic duct, which runs up through the chest and empties into the jugular vein at the base of the neck.

The blood may be noticeably milky after a heavy meal of fat. The rest of the products of digestion are passed into the capillaries, and dissolved in the blood. All the veins from the gut run to the liver, and here the food is further dealt with. Thus sugar, if not needed immediately for oxidation, is stored as a starchy body called glycogen, discovered by Claude Bernard, and gradually liberated as required later on.

Some of the sugar and most of the fat are stored elsewhere. Part of the sugar is stored as glycogen in muscles, but much of it is made into fat and stored under the skin and round some of the internal organs, along with fat from the food. In the frog fat is stored in special fat-bodies. The stored glycogen is later split up into glucose for use in the body by an enzyme, which continues to work after death, and the sugar thus liberated gives liver its well-known sweet taste. Absolutely fresh liver is not sweet. Again, ammonia, which is formed in the digestion of many proteins, is, in the liver, mostly combined with carbon dioxide to form urea ($2\ NH_3 + CO_2 = CON_2H_4 + H_2O$).

Urea is a very innocuous body, but ammonia rather poisonous if it gets to the brain, so that if the blood from the gut is short-circuited into the vena cava instead of going into the liver, a heavy meal of meat may cause convulsions. The liver also deals on the same sort of lines with any excess of amino-acids in the blood and with various poisons; its most important function is thus the regulation of the blood's composition, and not the secretion of bile. During starvation the body lives on its stores of fat, and the liver takes on the new duty of converting the stable fats such as those of suet into oils like that of linseed, which are very easily oxidized. The unabsorbed residue of the food from the small intestine passes into the large intestine, where it remains in man for a day or so, and is acted on by bacteria.

These are of little or no value to man, though very valuable to some animals. In man little but water is absorbed there. By this removal of water the bulk of the waste food is reduced by about nine-tenths. The large intestine also excretes poorly soluble salts, such as calcium phosphate, from the blood. These would clog up the urinary passages if they were excreted by the kidneys. The active tissues take oxygen, sugar, fat, and aminoacids from the blood, and use them for oxidation, growth, and repair. Into the blood they empty waste-products, of which the most important are water, carbon dioxide, and urea, but other soluble waste-products of protein metabolism include sulphuric and phosphoric acids, creatinine, and uric acid.

The last two contain C, H, O, and N. They are all excreted by the kidney except the carbon dioxide and some of the water, which go out by the lungs, and, in the case of water, the skin. The kidneys consist of a mass of tubules (about a million on each side in man) each beginning in a capsule containing a tuft of capillaries, and ending, after a winding course, in the central cavity of the kidney.

The capillary tuft seems to act as a filter, and the fluid that soaks through it is blood minus corpuscles and minus a few of the large molecules, such as the proteins concerned in clotting. As this filtrate runs down the tubules, the cells lining them reabsorb valuable constituents of the blood, such as sugar, and probably add unwanted ones such as urea, ammonia, uric acid, creatinine, and sulphates. The urine trickles down the ureters into the bladder, whence it is emptied from time to time.

An adult man produces on an average about 1.5 litres of urine per day, containing 30 grams of urea, 15 of sodium chloride, and 10 of other soluble waste-products. The blood is thus the medium of exchange between the different parts of the body. The heart keeps it moving, the lungs and gut supply it with fresh oxygen and foodstuffs, other organs get rid of waste products, and all the tissues of the body take from it according to their needs. The life of every part of the organism therefore depends on an adequate supply of blood of proper composition.

THE NERVOUS SYSTEM

The activities so far described may almost all be found in machines, but a complicated machine needs human control if the parts are to work harmoniously. The body to a large extent runs itself. The diaphragm and heart contract with the appropriate force and rhythm, the pancreas begins to secrete as food reaches the duodenum, and so on. Many of its activities are unconscious: consciousness and will only play a part where our past experience is likely to be of value in influencing our behaviour.

The unconscious responses of the nervous system are called reflex actions, or reflexes; though of course it is also responsible for voluntary actions. The nervous system controls striped muscle, heart muscle, smooth muscle, and glands, but with very few exceptions it is only the striped muscles that the will can influence, and even they are often moved by reflexes.

In every reflex or voluntary action three organs are always concerned, first a receptor organ which is appropriately stimulated, then a longer or shorter path in the nervous system, and finally an effector organ. The latter is always a muscle or gland in man, though other animals have electric and luminous organs under nervous control.

In the case of voluntary action the delay in the central nervous system may be very long, but there is always some external motive for a voluntary action. The nature of reflex and voluntary action will be made clearer by a few examples.

	Action.	*Receptor.*	*Effector.*
1.	Speeding up of the heart on increasing its supply of blood.	Nerve-endings in right auricle.	Heart muscle.
2.	Contraction of pupil in strong light.	Retina of eye.	Smooth muscle of iris.
3.	Reddening of skin after a scratch.	Pain spots of skin.	Smooth muscle of small vessels which open.
4.	Secretion of saliva on smelling food.	Olfactory organ in nose.	Salivary glands.
5.	Knee jerk.	Nerve-endings in tendon.	Extensor muscles of thigh.
6.	Blinking on eye being struck at.	Retina of eye.	Eyelid muscles.
7.	Breathing.	Respiratory centres in brain.	Muscles of chest and diaphragm.
8.	Sneezing.	Nerve-endings in nose.	Muscles of chest and diaphragm.
9.	Answering a bell.	Organ of Corti in ear.	Leg muscles.

All but the last are reflexes. The first three are entirely independent of will or consciousness. They are performed by smooth muscle. The fourth involves consciousness but not will. It can, however, be influenced by voluntary attention. The next four are performed by striped muscle, and are partly under voluntary control. The last is a very simple voluntary action.

The line between reflex and voluntary action is not sharp. Only an experienced schoolmaster can tell voluntary from reflex coughing.

It will be seen that the receptor organs are generally, but not always, sense-organs, that is to say, their stimulation produces consciousness as well as reflex action. We can learn a great deal about the properties of nerve by taking a muscle with its motor nerve out of a recently killed animal. If we stimulate the nerve by electrical or chemical means, or mechanically (e. g. by pinching), the muscle will contract. This irritability continues for many hours, though the muscle is very easily fatigued unless it has a proper oxygen supply. The muscle may be made to work a lever which writes on a moving sheet of paper, and the effects of different stimuli can thus be compared.

We can also measure the heat or electrical changes produced. By such means we learn the following facts about nervous conduction. Each fibre conducts independently of the others. It conducts not a steady stream, but a series of nervous impulses. An impulse is not an electric current, but an activity of the nerve-fibre producing an electrical effect and a little heat as it goes along. It travels at about 30 metres per second or 70 miles per hour in man.

Thus a man, hit by a car going at 80 miles per hour, will probably feel nothing because his brain is destroyed before any nervous impulses from his skin reach it. After the passage of an impulse the fibre needs a rest of one-thousandth of a second or more before it can transmit another.

All impulses in the same fibre are normally of the same intensity. Thus, from this point of view, we may compare a nerve to a bundle of telegraph wires, down which electrical waves of the same intensity pass at varying intervals, but not to a bundle of telephone wires, in which the intensity of the waves is variable. Finally, the energy of a nervous impulse is so small that about four million impulses (which would take several hours to pass) would be needed to heat a nerve 1° C. A voluntary muscle responds to one nervous impulse by a twitch, to a rapid series by a steady contraction.

A contracting human muscle is getting about forty-five impulses per second. The energy liberated in a gram of contracting muscle is several hundred thousand times greater than that in a gram of the nerve which supplies it. This ratio is about the same as that of the energy developed by a thirty H.P. motor-car running for twelve hours, to that used by the man who turns the starting-handle for a minute.

Muscles can be made to contract by weak artificial electric currents as easily as by those produced by the nerve. We are not yet sure of the details of how a muscle contracts, though it seems that the lactic acid formed causes microscopic fibrils to contract, as many proteins do when placed in weak acid.

A muscle is not a heat engine, for it has a very high efficiency such as is only found in heat engines one part of which is very much hotter than the other. Its chemical energy is converted directly into work without first passing into heat. The actual process of contraction may have an efficiency of 90 to 100 per cent., but an amount of energy greater than the work done in contraction is wasted as heat during the re-synthesis of the lactic and phosphoric acids, so that the whole process has an efficiency of only about 40 per cent.

Moreover, a good deal of energy is needed for the extra breathing and heart action during exercise, besides the basal metabolism which goes on all the time. So, considered as a machine, a man never has an efficiency of more than 25 per cent. At best, he turns three times as much energy into heat as into work. Moreover, a muscle heats up while keeping up a steady contraction, as in standing, supporting a weight, or pushing at a closed door. Striped muscles become quite flabby when their nerves are cut, but heart muscle and smooth muscle remain active.

Striped muscles have probably only one set of motor nerves which make them contract. Involuntary muscles have two sets: stimulation of one set causes increased activity, of the other set rest or lessened activity. The effects of stimulating a nerve-fibre depend mainly on its connexions in the body, to some extent on the quantity and rhythm of the stimulus, but not at all on where in its course it is stimulated. These facts were first discovered by Müller in 1826. Thus a blow on the 'funny-bone' or just above it is felt in the ring and little fingers because the nerve from them runs near the surface at this point, and irritation of the nerves in the stump of an amputated leg will give a man pain which he feels in toes that he may have lost forty years ago.

Again, when the nerve to the face muscles has been destroyed, the power to move them may sometimes be regained by grafting the nerve supplying certain neck and shoulder muscles into the old track of the facial. But when connexion has been made the patient, in order to move the face, must will to move the shoulder. The most accessible receptor organs are those of the skin.

They can easily be studied in an area where they are scattered, as on the side of the knee. If the skin is shaved we find that only parts round the hair roots are sensitive to gentle touch with a bristle.

Each root is surrounded by a network of nerve-fibres which are easily stimulated. The small hairs act as levers, and render the 'touch spots' more sensitive. In hairless parts, such as the palm, sole, and lips, there are special receptors for touch. Similarly if we go over the skin with a warm blunt metal point, we find that warmth is only felt at a second set of points (not those sensitive to touch), cold at another set, and pain at a fourth.

Many areas on the thigh are quite insensitive to pain, as they have no pain-spots. It is characteristic of receptor organs to be specially sensitive to one kind of stimulus, which may be physical, as with the skin organs, or chemical as with those of taste and smell.

They will, however, generally respond to inappropriate stimuli, if these are strong enough. Thus mustard will stimulate first the heat and then the pain-spots. A blow or pressure on the eye will make one see stars, and so on. On the other hand each receptor organ, with the paths leading from it to the brain, can generally only give rise to one kind of sensation.

This, however, depends on the part of the brain to which it leads, not on the organ which is stimulated. There is no fundamental difference in the nature of the impulses in different nerves, as there is in their effects.

If we stimulate the optic nerve, even after the loss of the eye, we get visual sensations, and so on. If we put the right hand into hot water, and the left into cold for a minute, and then both into lukewarm water, this feels cold to the right hand and hot to the left.

This is characteristic of the senses. They tell us more about differences of intensity in their stimuli than about their absolute intensity. When we look at a candle in sunlight we find it hard to believe that we can see by it, or even be dazzled by it, at night. We go into a dark room, and see nothing at first, but soon adapt ourselves, and see the things in it instead of a uniform blackness.

After half an hour in a sound-proof room one finds the noise of one's own heart and breathing unpleasantly loud, though normally one cannot hear them. The fineness of discrimination for touch depends mainly on the closeness of touch-spots. Thus, on the palm, where they are very numerous, we can distinguish two points from one if they are 1 centimetre apart. On the back, where there are few touch-spots, this distance must be increased seven times or more. Under the skin are receptors of many kinds. Some respond to deep pressure, and others to pain, but when once the skin is cut through, most healthy tissues are almost insensitive to pain. They become tender, however, when inflamed.

The gut and other hollow organs are insensitive to cutting or burning (stimuli to which they are not normally exposed), but very painful when stretched either by unusually bulky contents or unusually strong contractions of their muscles. There are also receptors in muscles, tendons, and joints. These send impulses to the central nervous system, which inform it of the relative positions and movements of different parts of our body.

Receptors and nerves with this function are called proprioceptive, whilst those whose stimuli come from outside are called exteroceptive. Proprioceptive organs may affect the consciousness. Thus we can tell how much our knee is bent even with our eyes shut, owing to the jointorgans, or how great a weight we are holding, owing to the muscle-organs.

But far more important is the aid they give us without our knowing it, in the co-ordination of muscular movement. If proprioceptive impulses cannot reach the brain from the legs, as happens in a disease of the spinal cord called locomotor ataxy, in which sensation is not lost, the patient cannot co-ordinate the movements or postures of his leg muscles.

In walking he raises his foot too high and brings it down too hard. He cannot stand with his eyes shut. There is no weakness of the muscles, but they cannot be used properly, as the brain gets no information as to what they are doing except through the eyes. The spinal cord consists of a central core of nerve-cells, called the grey matter, surrounded by millions of fibres mostly running lengthways, and called the white matter.

Both include a scaffolding of supporting cells. Between each pair of vertebrae a nerve leaves the spinal canal on each side. It enters the spinal cord by a dorsal and a ventral root. The ventral root fibres go to muscles and glands, and are only traversed by impulses going outwards. The dorsal root consists of fibres carrying impulses from receptor organs to the cord. So if a dorsal root is cut we lose the capacity for feeling with a certain area of the skin, while injury to ventral roots leads to paralysis of muscles. But most nerves contain both sensory and motor fibres, so when a nerve is cut both movement and sensation are lost in the area which it supplies.

The separation of the roots serves to bring all the sensory fibres to one cell-area within the cord, all the motor fibres to another. A nerve's only function is to conduct, but the spinal cord not only conducts impulses to and from the brain with its fibres, but gives rise to reflexes by means of its nerve-cells.

This is shown by what happens when it is divided. If a man breaks his spinal cord in the neck, he dies because his breathing muscles are cut off from the brain, and get no nervous impulses to make them work. If it is broken lower down he may live for some time. He has absolutely no feeling in the parts of his body and no voluntary control over the muscles whose nerve-supply comes from the part of the cord below the break.

But if we examine him six months after the accident we find reflexes occurring in the lower part of his body. If, for instance, we pinch his foot it is drawn upwards without his knowledge or will. If the lower part of the cord is destroyed or the nerves to it cut, all reflexes cease. A great deal has been learnt about nervous activity from the study of spinal reflexes.

They are easily studied on the carcass of a frog whose brain has been destroyed by poking a blunt wire into it from behind. If we irritate its skin its hind leg scratches near the place irritated, but its responses are clumsy, and it does nothing without some fairly violent stimulus. Provided they are in nervous connexion with the brain or spinal cord, the muscles of a limb are never quite flabby. They are mostly in a state of gentle but steady contraction or 'tone', so as to keep the limb in a definite posture.

If then the knee is to be bent, as the flexor (bending) muscles of the thigh contract, its extensor muscles must relax. If they relaxed too little or too slowly there would be a strain and a waste of energy. If they relaxed too quickly or completely the movement would proceed too far, and the kneejoint might be dislocated. Exactly the same applies elsewhere. Almost every muscle in the body, those of the trunk, jaws, and eyes, as well as the limbs, has an antagonist, and arrangements must be made for one to relax as the other contracts.

As there are no inhibitory nerves to striped muscles, this can only be done by inhibiting or switching off the activity of those cells in the central nervous system which are sending impulses to the muscle which has to relax. Figure gives an idea of some of the connexions concerned in a simple spinal reflex. An impulse enters the cord through a fibre in a dorsal root from a pain-spot in the foot.

The fibre divides and its branches end near nerve-cells in the grey matter. One of these cells is represented sending a fibre to a flexor muscle of the knee, another sends a fibre to an extensor. When the pain-spot is stimulated the impulses passing along it cause more nervous impulses to be generated in the cell connected with the flexor muscle, less in that connected with the extensor, so the knee tends to bend, and the foot to be withdrawn. Actually things are far more complicated. Stimulation of a single pain-spot will only cause movement after a long time or never, and, if movement occurs, hundreds of nerve-fibres will be conducting impulses at once.

To get a prompt movement one must stimulate a number of spots or fibres from them at once, as when one treads on a hot coal. Though the type of connexion shown in the figure have actually been observed with the microscope, in most reflexes the excitation has to pass through several neurons before it reaches the cell whose axon is the nerve-fibre to the muscle or other effector.

We must now study the function of the spinal cord in conducting nervous impulses in both directions between the body and brain. In a mixed nerve like the sciatic in the thigh, all sorts of fibres with different functions run together. Some are carrying impulses to the muscles, others from the skin and deep receptors.

As they enter the cord they are sorted out according to which way they conduct, and later on a further sorting process occurs. For example, the impulses which on reaching the brain give rise to sensations of temperature, run up the spinal cord by a different path from those which give rise to sensations of touch.

These paths have been located by several different methods, which have also been applied to the study of the brain itself. First, symptoms are observed in patients, and after their death local injuries of the spinal cord due to splinters of metal or bone, burst blood-vessels, or tumours, are found. When the same symptoms are observed in another patient they can often be relieved by the surgeon owing to the knowledge so gained.

To refuse leave to examine the body in such a case is to condemn some one else to die with those symptoms. Again, after destruction or division of some parts of the nervous system one can observe with the microscope the death and degeneration of groups of nerve-fibres. Now we know that when a fibre is divided, only that part dies which is separated from the cell-body and nucleus of the neuron to which it belongs. So we can discover in which direction the cell-bodies of any bundle of fibres lie. But in the central nervous system the long fibres always conduct nervous impulses away from the nucleus, so we discover the direction in which nervous impulses run in the fibres we have cut.

Finally, we can try the effect on an animal of cutting or stimulating some part of its nervous system. (The lower parts of the system will still work after the animal has been made unconscious by an anaesthetic or by removing its cerebrum). By such methods we can distinguish five main pairs of ascending fibre-tracts in the human spinal cord, besides numerous smaller groups. Two of these go to the cerebellum, and their injury does not affect consciousness, but causes unsatisfactory movements and postures, the brain being without information as to what the muscles are doing.

Two of them send impulses only to parts of the brain concerned in consciousness. One serves both purposes. Before we study the functions of the brain it will be convenient to deal with the special sense-organs in the head which communicate with it directly and not through the cord. The organs of the chemical senses, taste and smell, are found in the mouth and nose. They work together, and much of the sensation we commonly regard as taste includes an element of smell. With the eyes and nose tightly shut, taste will not distinguish an onion from an apple.

The tasteorgans mostly lie in the papillae which roughen the upper surface of the tongue. There are four elementary kinds of taste, namely: salt, sweet, sour, and bitter. Other tastes are combinations of these. Each elementary taste has different end-organs.

Thus, we taste sweet things best with the tip of the tongue, bitter with the back. The end-organs of smell are a little patch of about onequarter of a square inch of yellow epithelium at the top of the internal cavity of the nose. The corresponding area in a dog is ten or more square inches, in a large shark 24 square feet. In man, smell is an unimportant, almost vestigial sense, but in the dog and many other animals it is the most important of all.

So the dog's world is mainly a world of smells. But even in man it is the most delicate of the senses. We can smell mercaptan at a dilution of 1 milligram in 20,000,000 litres of air.

As about 1 cubic centimetre at a time is in the olfactory part of the nose, this means that we are affected by one twenty-thousand-millionth of a milligram, whereas the smallest object we can see with the naked eye is about a million times as large. In ordinary breathing most of the air goes past the olfactory cavity. In sniffing, some is sucked into it. No satisfactory classification of smells has yet been made.

Chapter 8

Organic Regulation

The nervous system serves to co-ordinate the activities of the different organs to some extent, but it is not in itself essential for the life of the tissues. A leg will live for years without nerves, but only for an hour or less without blood or some artificial substitute for blood. The cells in a higher animal are like skilled workmen, very efficient at their own job, but not at other jobs.

Thus a single cell in Hydra may serve for protection, be sensitive to external stimuli, contract when stimulated, pass on excitation to its neighbours, and perhaps secrete mucus, but in none of these ways will it act as efficiently as the various cells of a mammal, each of which performs one of these special functions.

The latter are enabled to specialize largely because they have a nearly constant environment, constant in chemical composition and temperature, and do not spend any energy in adapting themselves to change in it.

This environment is supplied by the fluid part of the blood. In this and the next chapter we shall consider some of the factors in the internal environment, and how they are kept steady or adapted to new conditions.

Most of the general symptoms of disease are due to upsets of the internal environment. Each organ must have food, oxygen, and a means of getting rid of waste products. But this is not all. It must have them in the right amounts. Too much oxygen is just as deadly as too little. And it must also have in the right amounts other substances which it does not use for work or repair. An animal dies if we halve or double the amount of potassium salts in its plasma, though it does not turn potassium salts into anything else.

So these too have to be kept steady. Further, an organ may need different amounts of a given substance at different times. If a muscle suddenly starts work, its O_2 -consumption and CO_2 -production increase about fifty times. It was already using most of the oxygen in the blood which passed through it, so it must increase its blood-supply correspondingly. When it is at rest about five out of every six of its capillaries are shut. When it begins to contract these at once open and the others open wider. The small arteries also open wider. In a resting muscle they are kept almost shut by the slight alkalinity of the blood, and also probably by the presence of oxygen.

Now if we cut the leg off a recently killed frog and run fluid through the blood-vessels, they will open up if this fluid is not sufficiently alkaline, or is short of oxygen. Just the same thing happens when a muscle contracts or a gland begins to secrete.

The O_2 flows from the blood and CO_2 is poured into it, making it acid; the blood-vessels relax and widen, and the organ obtains an adequate flow of blood. Other products of activity besides CO_2 probably co-operate in producing this effect. But when any large organ opens up its vessels the arterial pressure would fall unless the heart pumped harder. The other organs would then go short of blood. We must therefore study the working of the heart.

Like other involuntary muscles, it will work without any nervous control. If we take the heart out of a recently dead animal or man, and supply it with warm blood or an appropriate salt solution containing oxygen, it begins to beat again and may go on for many hours. Even an isolated piece of it will beat. In a mammal the beat starts at the entrance of the great veins to the right auricle in a special piece of tissue known as the 'pacemaker', which does not contract but stimulates the neighbouring muscle.

If we warm the pacemaker, the whole heart beats faster; if we destroy it, the heart first stops, then begins to beat at a slower pace of its own. The auricles contract almost instantaneously when stimulated by the pacemaker, but they are only connected with the ventricles by a narrow bridge of conducting tissue in which the wave of excitation is delayed for about one-tenth of a second, and then passed on almost simultaneously to all parts of the ventricles.

If the bridge is damaged, as in some forms of heart-disease, the ventricle may only respond to every second or third beat of the auricles. If it is destroyed they beat at their own rather slow rate, and cannot be speeded up. Now if we take an isolated heart, or a heart whose nerves have been cut, and give it an increased supply of blood, it can increase its output per beat but will not increase its rate at all.

The rate is governed by two pairs of nerves. Of these, the vagi are one; if they are stimulated the heart slows down; if they are cut it speeds up, showing that they are normally acting as a brake on it, mainly through the pacemaker. The other pair, called the accelerators, come through sympathetic ganglia from the spinal cord.

Stimulation of them speeds up the heart. Both pairs are governed by the same centres in the medulla oblongata. If the blood-supply to the heart is increased, the great veins and auricles are distended, and receptor organs in their walls send impulses up to the brain which result in the vagus brake being slackened by a reflex action; and if the stimulus is sufficient, the accelerators are set to work.

Hence, when more blood reaches the heart from the open vessels of an active muscle, it increases its rate and force. Another set of reflexes keep the arterial pressure steady. A pair of nerves called the depressors run from receptor organs in the aorta to the medulla oblongata, which they enter with the vagus. If the aorta be distended by an abnormally high blood-pressure, impulses run up them to the medulla. The reflex response to this is a slowing of the heart by the vagus and an opening of small arteries. The opposite occurs if the aortic pressure falls. There is also a pressure-gauge in the brain itself.

If pressure is put on the brain from outside, for example, by a clot of blood under the skull, its vessels will collapse, and the brain will force the heart to raise the arterial pressure till they open up again. In this way the arterial blood-pressure is kept steady, so that any organ can obtain the blood-supply it needs by opening up its bloodvessels. But the brain does not allow an indiscriminate competition between different organs for blood-supply.

The arteries as well as the heart are under nervous control. A series of nerves called vasoconstrictors run in the sympathetic system to the smooth muscle of the arterial walls. The vasoconstrictors come into play in circumstances which affect the body as a whole, such as change of posture, or violent exercise.

When a man gets up after lying down the blood tends to flow into his belly and legs, and this is prevented by the contraction of the arteries of these parts, under impulses from the vasomotor centre, which lies near the heartregulating centre in the brain. If he has been in bed some days the vasomotor centre is out of practice and his brain runs short of blood, so that he becomes *dizzy* and may faint. If there is a great deal of blood in the guts and skin, as when we are sitting before the fire after a heavy meal, this may even happen on getting out of a chair. Again, during muscular exertion the arteries to the guts contract, and digestion has to stop.

The vasoconstrictor nerves are also excited by chemical stimuli to the brain. If we are throttled or breathe very impure air, the CO_2 of the blood goes up and the O_2 down. The vasomotor centre then narrows down all arteries except those to the heart, lungs, and brain, and the blood-pressure rises. These three essential organs must have an adequate supply of oxygen, whatever else goes short. The other organs, if left to themselves, would open up their vessels, but in a general emergency they are not allowed to do so. The heart is also speeded up. There are also vasodilator nerves.

For example, when a dog gets hot the vessels in its tongue are opened up by a special nerve. Moreover, many of the nerve-fibres whose stimulation causes pain send branches to the local blood-vessels. When a pain-spot is stimulated, impulses run up to the spinal cord. They also run directly to the local vessels, which open up, causing reddening of the skin;

this is almost the only reflex in higher animals of which the nervous path is entirely outside the central nervous system. The above are cases where an organ gets blood which is not to be used mainly as a gas-carrier. The same occurs in an actively secreting gland. A salivary gland when active uses three times as much oxygen as when at rest.

But it also needs a great deal of water to make saliva, and it is as a source of water rather than of oxygen that it needs blood. Its bloodsupply must go up five times, and as the ordinary chemical call for blood is not effective, a vasodilator nerve is used. As the blood-supply goes up more than the oxygen-consumption, the venous blood of the active gland is actually redder than usual. A much more important case of non-respiratory bloodsupply is the skin's. The human skin excretes energy just as the kidneys excrete matter. All the heat produced in the body has to get out, and seldom does more than a fifth of it get out in the breath.

The remainder goes through the skin, and its loss is regulated in such a way as to keep the temperature of the body very constant. If we go into a hot room the same amount of heat has to be lost in a given time, but it is obviously harder to get rid of it.

If we work hard and produce more heat in our bodies, more heat has to be lost in a given time, though the loss of a given amount is no harder. In each case the skin responds in the same way. Its small arteries open up and it gets red and warm. Heat is thus brought from the inside to the surface in large amounts and rapidly lost, as from the radiator of a motor vehicle. If this means of losing heat is insufficient, we begin to sweat.

The sweat comes from microscopic glands under nervous control. It consists of water containing less salt than the plasma. When this evaporates the skin is greatly cooled, for water has a big latent heat of evaporation. Sweat that does not evaporate does not cool us, and it cannot evaporate if the air is already saturated with water-vapour. So sweating is useless in a hot and damp atmosphere, which is therefore far more oppressive than a dry one of the same temperature. The ordinary thermometer does not tell us whether we shall be able to lose heat or not.

For this purpose we use a 'wetbulb' thermometer. The bulb is wrapped in a wet cloth, so that the drier the air is, the more heat it can lose by evaporation. It is therefore in the same position as a man whose clothes are soaked with sweat. If the air is saturated with water the wet- and dry-bulb thermometers have the same reading, if the air is dry the wet-bulb thermometer may read more than 100° F. lower. Men can stand dry heat far above boiling-point, staying in a room where a steak is cooked in five minutes, and only coming out when their hair begins to singe.

But a wet-bulb temperature above 90° F. is fatal, and above one of 75° F. the capacity for work is lowered. We get an idea of the efficiency of sweating by considering a man in fairly dry air at body-temperature (98·5° F.). He can lose no heat by conduction or convection, so it must all be used in evaporating water.

He has to lose 3,000 kilocalories per day. But the evaporation of 1 litre of water at body-temperature requires 570 kilocalories, so in a day he must sweat 5·3 litres, or 9·3 pints. Actually many men can sweat 1 litre per hour, and the world's sweating record is held by an English coalminer who lost 18 pounds (1·8 gallons, or 8 litres) in 51/2 hours. To make up for the loss of sweat one must drink more water and eat more sodium chloride than usual.

Miners working in great heat are therefore fonder than the average man of bacon, kippers, and salt. Many animals, such as dogs, have very few sweat-glands, but produce a very thin saliva which they evaporate by rapid shallow breathing, panting with open mouth and tongue hanging out.

Several parts of the brain are concerned in heat regulation. They receive nervous impulses from the skin, and also from local organs in the brain which measure the temperature of the blood like thermometers. Thus, if certain parts of an animal's brain, or the blood going to them, are heated, the animal begins to flush and sweat, and the rest of its body is cooled down.

If the brain is cooled the animal shivers and its temperature rises. During adaptation to heat we cannot cut down our heatproduction except by keeping still, but when cold, besides shutting down the skin-vessels, we first tighten up our muscles, then shiver, and finally take exercise. In all these ways more heat is produced. In many diseases the temperature rises.

This is not due to increased heat-production, but to diminished heat-loss owing to perverted function of the temperature centres. A man whose temperature is thus rising feels very cold until it has reached the new level to which he is regulating.

He shivers and complains of the draught, to which he may put down his illness. If his temperature falls quickly he sweats profusely and feels very hot. Mammals and birds have a nearly constant temperature, but other animals have a variable temperature, a fraction of a degree above their surroundings. If we warm a 'cold-blooded' animal through about 5° C. we double its rate of oxygen-consumption, and all its other activities.

For example, it is possible to read the temperature within 1° F. by measuring the distance walked by an ant in a minute! Cold-blooded animals cannot move quickly in winter, and mostly die or rest in holes. The activities of mammals and birds are not slowed down, so they are the dominant animals in temperate and cold climates. But in hot countries, snakes, crocodiles, and so on, are able to compete successfully with warm-blooded animals.

A few mammals, such as hedgehogs and dormice, compromise by sleeping through the winter at a low temperature (and therefore a low rate of oxidation), but they never let their temperature fall to that of their surroundings. We must now turn to chemical regulation of the composition of the blood and tissues.

It will be convenient to begin with the gases, the quantity of which in the blood is regulated by breathing. The obvious duties of the lungs are to get rid of CO_2 and let in O_2, and if we go into a room containing say 6 per cent. of CO_2 instead of the normal 0·03 per cent., or 10 per cent. of O_2 instead øf the normal 20·9 per cent., the breathing increases greatly. However, a small drop in the O_2 of the air breathed has no visible effect on the breathing, because the haemoglobin is already almost saturated with oxygen at a pressure less than that in the lungs. So want of O_2 cannot be what normally keeps the breathing going.

To find out how the breathing is regulated we must take the samples of air from the very bottom of the lungs, where it is in equilibrium with the blood. This, which is called the alveolar air, can be obtained at the end of a deep breath out. The amount of carbon dioxide in it is very constant, about 51/2 per cent., whereas the amount of oxygen varies a good deal. If the amount of carbon dioxide increases by only 3 per cent. of its normal value, the breathing is doubled; if it falls by the same amount, as after voluntary over-breathing, the breathing stops.

The main reason why we breathe more during moderate muscular exercise is because more carbon dioxide is being produced, and this stimulates the respiratory centres in the brain to make the breathing muscles do more work. Thus the lungs have the function of keeping the CO_2 pressure in the tissues at the normal level, not merely of excreting it. Carbonic acid seems merely to act on the respiratory centre in virtue of its being an acid. If another acid, such as hydrochloric, is injected or drunk, the breathing is greatly increased, while it slows down when an alkaline substance such as sodium hydrogen carbonate is taken. The most familiar case, however, is that of very violent exercise.

When the muscles are working so fast that they cannot get enough oxygen for their recovery process, lactic acid accumulates in them and leaks out into the blood, from which it is only gradually removed. So after running a quarter-mile the extra carbon dioxide is got rid of in the few minutes of violent panting which succeed the race, but a small increase of the breathing, due to lactic acid, may persist for half an hour or so. During this time the alveolar carbon dioxide pressure is kept below normal by the extra breathing, thus compensating for the acidity which would otherwise be produced by the lactic acid. Serious oxygen-want also excites the respiratory centre.

If one goes into air containing only about half the normal 20·9 per cent. of O_2, one at once begins to pant, but after a while the panting dies down, because a lot of CO_2 has been blown out of the body by the increased breathing, and the respiratory centres have no more reason to discharge nervous impulses than before, their normal stimulus, CO_2, being reduced in quantity. So a man who has gone into bad air at first pants enough to keep his blood supplied with oxygen. Then the breathing becomes normal, and he falls unconscious with oxygen-want.

A candle is often a much better measure of oxygen-want than one's own feelings. Another way in which the breathing is affected is by the process of digestive secretion. When the stomach secretes hydrochloric acid the blood would be too alkaline if carbonic acid were not kept back to take its place, so the breathing is slightly slowed down. Later on, the pancreas and intestine begin to remove alkali from the blood for their secretions, and to prevent it getting too acid the breathing has to be increased.

These changes are too small to observe directly, but can easily be measured. We can get some idea of why the alkalinity of the tissues has to be regulated so carefully by experimenting with tissue or enzymes taken from them. If we take a dead organ and preserve it carefully from bacteria it does not putrefy. But if it is kept at body-temperature the tissues gradually soften and are found to be digesting themselves.

This is due to enzymes in them. The dying tissues produce acids, and in a slightly acid medium these enzymes work very much more rapidly than in an alkaline or neutral one. Thus, to prevent an organ from digesting itself it must be kept slightly alkaline. The best known of these enzymes is that in the liver which breaks up its glycogen into sugar. A quite fresh liver, besides being very tough, does not taste sweet. If it is allowed to digest itself for a few days it not only becomes tender, but sweet.

Some other enzymes act more rapidly in a medium more alkaline than the normal, so if the reaction of the tissues is altered the normal balance between the different chemical processes is upset, and death may occur. Just as the lungs regulate the amount of gases in the blood, the kidneys regulate the amount of the soluble bodies.

The blood which passes through these organs is always altered so as to resemble an 'ideal' blood. Thus, if there is more water in the blood-plasma than in this ideal or standard plasma the kidney secretes an unusually watery urine, and the plasma of the blood in the renal vein therefore contains less water than the arterial blood, and resembles the standard plasma more closely.

If, as is more usual, especially in hot weather, there is rather less water in the plasma than in the standard plasma, the kidney secretes a concentrated urine, so the blood leaving the kidney contains more water than that entering it. The substances found in blood and urine can be divided into two classes. The first class includes almost all foreign substances, for example, iodides, dyes, or foreign proteins injected into the blood. These are removed by the kidneys, however little there is in the blood. It also includes some very important waste-products, such as urea, the substance which contains most of the nitrogen resulting from protein oxidation.

The rate at which such substances are excreted is roughly proportional to the amount in a given volume of blood. The second class includes most of the normal constituents of plasma, such as sodium, potassium, calcium, magnesium, chloride, bicarbonate, phosphate, and sugar. These substances are only excreted if the quantity of one of them contained in a given volume of plasma exceeds a certain limit, called the 'threshold'.

For example, the amount of chloride in the plasma is generally a few per cent. above the threshold, and there are, therefore, chlorides. in normal urine. But if we drink a lot of water after violent sweating, the amount of chloride in the plasma falls below the threshold, and it disappears from the urine.

Normal blood contains about 0·10 per cent. of glucose, a simple sugar. If a healthy man takes 100 grams of glucose the amount in the blood rises to about 0·13 per cent., but none appears in the urine.

The threshold value for the kidney is about 0·17 per cent. If, therefore, the arrangements for storing sugar are out of order, as in diabetes, a dose of 100 grams will make the blood-sugar rise above 0·17 per cent., and sugar will appear in the urine. Besides excreting substances found in the blood, the kidney makes a few substances.

For example, sulphuric and phosphoric acids are made throughout the body by the oxidation of the proteins. The kidney has to get rid of these, but its cells and those of the urinary passages are damaged by strong acids. It therefore excretes the sulphuric acid not as such, but as ammonium sulphate, which is neutral in reaction. But there is not enough ammonia in the blood to furnish all that is required for this purpose.

Ammonia and ammonium salts are poisons when injected into the blood-stream, and the liver converts almost all the ammonia reaching it into urea, which is nearly harmless. So the kidney has to make its own ammonia, and the more acids it has to excrete the more ammonia it makes.

The kidney is doing work like a muscle, for work has to be done in concentrating substances, just as in compressing gases. For example, to concentrate the urea in a litre of blood into about 20 cubic centimetres of urine, as the kidney does every 20 minutes or so, requires at least as much work as to compress a litre of gas containing as many molecules as there are urea molecules in the blood, into 20 cubic centimetres.

Actually the number of urea molecules in a litre of blood is the same as that in a litre of gas at a tenth of an atmosphere pressure, so the work needed is that required to compress this gas to a pressure of five atmospheres, i. e. 40 kilogrammetres. In order to do this work the kidney needs oxygen. Its oxygen-consumption can be measured by determining the rate at which blood flows through it and the amount of oxygen lost by this blood.

As a matter of fact, the kidney uses a good deal more oxygen per gram per minute than the heart, and like the heart, will increase its oxygen-consumption three or four times if it is given work to do.

But if we inject salt-solution of about the composition of plasma, the kidney needs no more oxygen, although the volume of urine secreted per minute is increased. This is because it has no work to do in concentrating the salt, but it merely acts like a filter. If on the other hand we inject urea or sodium sulphate, its oxygen-consumption increases, as these substances have to be concentrated. Other glands behave in a similar manner, requiring more oxygen when stimulated to do work.

THE INTERNAL ENVIRONMENT

The importance of the extraordinary fact that an organ such as the heart or liver can be kept alive for many hours in a solution of inorganic salts which are present in nearly the same proportions as in plasma. For mammalian organs such a solution is:

NaCl, 0·8 per cent.; KCl, 0·02 per cent.; $CaCl_2$, 0·02 per cent.; $MgCl_2$, 0·01 per cent.; $NaHCO_3$, 0·1 per cent.

This solution, whose composition was worked out by Ringer, must be saturated with oxygen, and a little glucose may be added as a source of energy. Now if we leave out the various constituents, or increase them above the standard amounts, we shall find out what functions they are performing. The salts do not permeate the cells, or only do so very slowly, but affect their surface properties. If the solution perfusing, say, a rabbit's excised but still beating heart is diluted with water, the heart swells up and stops.

It has become sodden with water as does the skin of the hands in a hot bath, for water runs into the cells and dilutes their contents. If we make the solution too strong the cells shrivel up from loss of water. If we replace most of the sodium chloride by the corresponding number of molecules of cane-sugar the heart goes on beating. Clearly the main function of the sodium chloride is to prevent the cells from taking up too much water. If we leave out the potassium the heart goes into a state of cramp. Potassium is needed for its relaxation. If we leave out the calcium it stops in a flabby condition, and so on.

A heart so stopped may be revived hours later by adding the missing salt. Other organs behave in the same way. Thus, if there is too much or too little calcium in the fluid in its blood-vessels, the kidney refuses to hold back sugar when the amount of sugar in the fluid perfusing it is below the normal threshold value.

We can also observe effects of the same kind on a man or animal, though here the nervous system is generally affected before the other organs. Thus, if a man drinks water for some hours more rapidly than his kidneys can get rid of it he gets cramps, and later convulsions. If he lowers the calcium to half its normal value he gets another type of cramps, and so on. One of the most remarkable things about the plasma salts is that they are nearly the same as those of sea-water diluted with water to about three times its volume.

Such a solution would, however, among other things, contain too much magnesium and sulphate. Now the blood of marine invertebrates is very like sea-water, while that of sea-fish is generally somewhere between that of invertebrates and that of land vertebrates.

Many people, therefore, think that land vertebrates are descended from fish which left the sea when it contained less salt than now, and that our blood-plasma has kept the composition of the sea-water to which the cells of our ancestors were accustomed. Besides the inorganic substances mentioned, the plasma contains phosphates, which are also kept in by the kidney.

They play a very important part in the formation of bone, which consists largely of calcium phosphate. There is an exceptionally large amount of phosphate in the plasma of growing children, and of adults who have broken a bone and or of calcium in a child's plasma falls below normal it is unable to form bone properly, and develops rickets. Although an organ can be kept alive for many hours by inorganic substances, yet organic compounds are of course needed for its prolonged existence, and a heart will survive longer if, for example, a little sugar is added to its salt solution.

As we saw, the blood contains about one part in a thousand of sugar, and this does not fall much in a starved man or animal. In this case glycogen is broken down by the liver to keep the sugar-level up, and after the glycogen is mainly used up, the organs then oxidize fat rather than sugar, and leave the blood-sugar about normal. In the same way the amount of amino-acids remains fairly steady, though of course there is a small temporary rise after a protein meal.

The amount of fat varies somewhat more, and after a very heavy fat meal the plasma may be quite milky with microscopic fat-drops. The liver plays an important part in keeping the amounts of sugar and amino-acids steady. It stores extra sugar as glycogen, and removes the ammonia from excessive aminoacids.

The residue of the amino-acid molecule left can be oxidized, and in some cases can be made into sugar if required. But besides dealing with excess of normal bloodconstituents it can destroy poisonous substances coming to it from the gut. These substances are generally the result of bacterial action there. The bacteria attack the proteins of our food, but have not enough oxygen to utilize them fully. They therefore excrete unoxidized fragments of proteins, just as the yeast-cell excretes ordinary alcohol made from sugar which it cannot burn. Among these excretory products are phenol ('carbolic acid'), cresol, indol, and skatol. The last two are foul-smelling, and all are poisonous.

They are absorbed from the gut, but on reaching the liver they are combined with sulphuric acid, apparently by an enzyme, to form quite harmless substances which are excreted by the kidney. If, however, they enter the blood in very large amounts a proportion gets past the liver and the whole body appears to suffer. The liver acts in the same way with many other substances. It has more varied chemical functions than any other organ.

Besides foodstuffs and various colloids with which we shall deal later, the plasma contains in exceedingly small amounts certain organic substances of great importance which are poured into it by special organs, including the ductless glands, such as the thyroid.

Some of these substances, like adrenalin, are present in varying amounts and act as hormones or chemical messengers between different organs. Others seem to be present in fairly constant amounts, and are needed for the normal working of the body. Insulin is produced by certain microscopical 'islands' of tissue in the pancreas, and passes out of them into the bloodstream. If these islands of insulin-producing tissue are removed or diseased the tissues become more or less completely unable either to oxidize or store the sugar of the blood.

The amount present in the blood increases, especially after meals containing carbohydrates, and when it rises above the threshold value the kidney excretes it and it is wasted. What is even worse, the tissues begin to make sugar from proteins, and this too is excreted.

So a man with severe disease of the parts of the pancreas which make insulin gradually wastes away. This condition is called diabetes. If he is still making some insulin we can often keep him alive on a meagre diet which he can just deal with, but in severe cases we have every day, or several times a day, to inject insulin made from the pancreas of animals.

Unfortunately an overdose of insulin will depress the blood-sugar below normal, which may bring on convulsions and death. Insulin has not yet been obtained in a pure state, but we know that very little is needed, for less than one milligram per day of our strongest present preparation is needed even in severe diabetes. The thyroid gland in the neck makes a substance called thyroxin, which has been obtained in a pure crystalline form and whose chemical structure is known.

Absence or removal of the thyroid does not lead to immediate death, but to a condition called myxoedema in the adult mammal, cretinism in the young. In this state the resting O_2-consumption is only about 60 per cent. of the normal, and the adult patient becomes fat, sluggish, stupid, and bald; while a child develops abnormally and is idiotic; and a tadpole never metamorphoses into a frog.

These conditions can be completely cured by administering thyroxin or extracts of the gland. Fortunately thyroxin is not destroyed, like insulin, by digestive enzymes, so it can be given by the mouth. Thyroxin has such a powerful effect that only about a third of a milligram per day is needed to keep a thyroidless man normal, and this amount will cause the oxidation of a quarter of a million times its weight of glucose.

The thyroid gland sometimes swells up and produces too much thyroxin. The resting O_2 -consumption then rises and may reach double the normal amount. In one hospital the patients in the ward reserved for hyperthyroidism eat twice as much as those in any other ward! They become thin and nervous, and their eyes tend to protrude.

They can generally be cured by cutting out the gland wholly or in part. Thyroxin is an organic substance including four iodine atoms in its molecule, so if there is not enough iodine in the food and drink it cannot be made in sufficient quantity. In many inland districts swelling of the thyroid is found along with a low or normal resting metabolism. This disappears when a few milligrams a week of an iodide are given. The gland which has been over-working in an attempt to make thyroxin with too little iodine rapidly recovers.

The same simple remedy has made it possible to rear sheep in the American state of Michigan, where, owing to lack of iodine in the soil and water, they were formerly unable to live. On the other hand, healthy people may make too much thyroxin if given excess amounts of iodides. Another gland with important internal secretions is the pituitary, which produces at least two different substances. The posterior part produces a hormone, pituitrin, which affects smooth muscle. It seems to be needed for the normal function of the capillaries, which become relaxed and leaky in its absence, while the kidney produces large amounts of a watery urine.

It also has a powerful effect on the womb, and is used to help it to contract during child-birth. The anterior part of the pituitary is concerned in regulating growth. When it is removed we get a dwarfish condition, and when it is enlarged and secretes too much we get overgrowth. The precise effects depend on age. In a child, a long bone, such as the femur, consists of three parts, the shaft, and two bony pads at each end, called the epiphyses. Growth takes place not at the joints but at the soft cartilaginous junction between the shaft and the epiphyses.

If the pituitary begins to over-secrete before the epiphyses have been joined by bone to the shafts, the patient becomes a giant; if afterwards, he becomes a little taller, but the epiphyses, lower jaw, brows, and some other bones grow in thickness, and a condition called acromegaly results, characterized by large hands and feet and a peculiar facial expression.

Clearly for normal health the pituitary, like other glands, must produce just the right quantity of its internal secretions. The adrenal glands, which lie just above the kidneys, also consist of two parts. The central part secretes adrenalin, a substance which produces the same general effects as stimulating the sympathetic nerves; for example, a rise of blood pressure, a slowing down of the movements of the gut, and a pouring of sugar into the blood.

Adrenalin seems to be constantly entering the blood, but in larger quantities during emotion. The outer part (cortex) of the adrenals is necessary for life, and probably pours something into the blood, but we do not know what. The parathyroid glands, which lie in the neck in or near the thyroid, also produce an internal secretion which regulates the amount of calcium in the blood plasma.

The gonads also produce internal secretions, which control the appearance of the secondary sexual characters which distinguish the two sexes. For instance, they act on the nervous system, and cause the appearance of the instincts connected with sex.

They also influence the growth of the skeleton and other structures, such as combs and spurs in fowl, horns in deer, and the larynx and beard in man. At the time of writing one of their internal secretions has been obtained in a fairly active condition, though not yet pure. If we take a few living cells from an organ, and grow them under the best possible conditions, usually in plasma to which have been added extracts of embryonic tissue, they will continue to divide indefinitely if they are protected from bacteria.

Cells from a chicken's heart have grown in this way for over ten years, a few being transferred to fresh fluid every two or three days. It appears probable that they could be grown indefinitely in this way, so that in the body their death with the natural death of the fowl by old age must be due to causes outside themselves.

Blood contains substances which encourage cellular growth, and others which check it. As an animal grows up the former decrease in quantity. One cell may produce a substance which stimulates another type of cell. Thus a healing wound is full of leucocytes, which crawl into it from the blood-vessels.

As these die they liberate a stuff which causes another kind of cell, the fibroblast (which produces connective tissue), to grow and multiply, as can easily be shown on a tissue-culture. The fibroblasts in tissue near the wound therefore grow out into it, forming tough fibrous scar-tissue. Sometimes growth-regulation breaks down, and the cells of some part of the body grow too quickly, causing a tumour.

This may be harmless, like a wart, but if the growing cells migrate into the surrounding tissues and finally to distant parts of the body, the growth, which is then called a malignant tumour, or cancer, is very deadly unless it is removed when still small.

We do not yet know exactly why cells become cancerous, though in some cases this condition is undoubtedly caused by chronic irritation, for instance, by tar in tarworkers, by certain parasitic worms in rats. One of the most important functions of the blood is to clot. If this does not happen, as in certain diseases, a man may bleed to death from a tiny scratch. Blood does not clot in the blood-vessels, even if they are removed from the body, and clots only very slowly in well-greased receptacles.

But contact with broken cells or with most foreign bodies leads to clotting; in the latter case apparently because they cause certain of the formed elements in the blood to burst.

At least four substances are concerned in clotting. The clot is mostly made from fibrinogen, a protein dissolved in the plasma, but at least three other substances play a part, namely, calcium (lime salts), a protein called serozyme, and a waxy substance called cytozyme produced by the breaking-up of cells.

It is still uncertain exactly how they interact, but interference with any of them will stop the blood from clotting. For example, a little sodium or potassium oxalate will precipitate all the calcium of blood as calcium oxalate, and thus keep the blood fluid.

Of the other proteins in the plasma some are concerned in the defence of the body from infection, though it is possible that fat-like substances play a part too. Infectious diseases are caused by animals and plants, some of microscopic size, some, such as that which causes most common 'colds', too small to be seen with a microscope.

One of the best known of the visible ones is the bacillus which causes diphtheria. It grows in the throat, which it may occasionally obstruct, and does not usually spread to other parts of the body, but liberates into the blood a poisonous protein which is particularly dangerous to the heart. If a man has had diphtheria this protein, called diphtheria toxin, is no longer poisonous to him, and what is more, if some of his blood is mixed with the toxin, it renders it harmless when injected into some one else.

In practice one injects ground-up dead diphtheria bacilli into a horse so that it develops antitoxin without having had a sore throat. Its blood now contains antitoxin and will protect human beings against the toxin. We do not yet know the exact chemical nature of the toxin and antitoxin, but we know a great deal about them—for example, in what liquids they are soluble and at what temperatures destroyed. We also know that the antitoxin puts the toxin out of action by forming a rather loose compound with it.

Unfortunately, few bacteria kill us in the rather simple way employed by the diphtheria bacillus, so few diseases can be cured by antitoxins, although they are quite effective against snake-bite. Another type of immunity is that developed to typhoid bacilli. The serum of a man who has had typhoid, when added to a suspension of typhoid bacilli in water or salt solution, makes them stick together in little clumps and cease to move about.

The same or similar substances increase the capacity of the white corpuscles for digesting the bacteria, while others will actually cause foreign cells to break up. These substances are mostly very specific, and a serum which is fatal to one race of bacteria will not always kill a very similar race which produces a disease of the same kind, still less those which cause a different disease.

Besides such 'immune bodies' in the blood serum which can be transferred from one man or animal to another and experimented with in tubes, the cells of organs can develop immunity which, of course, cannot be transferred, and as to whose nature rather little is known. In many diseases most of the symptoms are due to changes in the blood-composition, which affect organs remote from the one which has first been injured. For example, in heart disease the first symptom is often faintness after exercise. This is because the damaged heart cannot pump blood fast enough through the lungs to absorb the oxygen needed by the body, so the brain does not get enough oxygen.

Many of the symptoms of lung-diseases such as pneumonia are also due to oxygen-want. When the kidneys are diseased they cannot get rid of substances with their usual ease, and they also let the proteins of the plasma leak out into the urine. Some victims of kidney-disease get headaches, vomiting, and convulsions as the result of the accumulation in the blood of various poisonous substances which a healthy kidney can get rid of, others swell up and get dropsy because the kidney cannot get rid of water and salts.

In most acute diseases the blood seems to contain poisonous bodies of unknown nature which upset the temperature-regulating centre, and so cause fever. Since bacteria and other living producers of disease were discovered we have been enabled to prevent many infectious diseases.

For example, we prevent the spread of typhoid and cholera by seeing that their bacilli do not get into drinking-water. Malarial fever is spread by mosquitoes, which suck up from the blood the protozoa that cause it, become infected themselves, and infect the next man they bite by means of their saliva, the protozoa having multiplied and migrated into the salivary glands.

Malaria can thus be prevented by killing off the mosquitoes. But, apart from surgery, rest, good diet, and nursing, we can often deal with disease by means of drugs. Among the most important drugs are antiseptics, which kill off bacteria in infected wounds or abscesses, and anaesthetics, which allow the surgeon to remove diseased organs painlessly. Some drugs, such as Epsom Salts, act in the gut, but most of them have to be absorbed into the blood before they act. One group of drugs are valuable because they are more poisonous to parasites than to human beings.

For example, a dose of quinine which will kill almost all the malarial parasites in the bloodstream only gives the patient a slight headache. But most drugs are valuable because they act on some special organ. For example, there is a form of heart-disease in which the auricles, instead of beating, twitch in an irregular way. The ventricles are constantly being excited by the auricles; they therefore contract so often that they have no time to fill up between beats, and become fatigued. We give the patient digitalis, a drug which acts mainly on the heart, and in some way blocks the conduction of impulses from the auricles to the ventricles.

The latter then begin to beat at a slow rhythm, but fill up properly between beats. Again, in chloroform or ether anaesthesia there is a danger of the heart stopping, so we want to block the passage of inhibitory impulses from the brain down the vagus. For this purpose we inject a little atropine. This has the property of stopping the action of the parasympathetic but not of other parts of the nervous system.

It not only hinders the action of the vagus on the heart, but also on the stomach, and this discourages vomiting. Many drugs act especially on the brain.

Some affect those parts which are concerned in consciousness, and produce sleep or wakefulness as the case may be. Others affect specific centres, such as the heat-regulating or vomiting centres, and by a suitable use of them we can allay pain, fever, and other symptoms which involve the activity of the brain. It is, however, certain that many more chemical weapons in the war against disease remain to be discovered.

To sum up, not only the normal activities of the body, but even its normal shape, depend on the simultaneous normal activities of its various parts. They influence one another partly through the nervous system and partly by mechanical means, as when a bone grows in response to the pull of a muscle, but very largely by chemical factors.

The chemical side of the relationship between the hundred million million or so cells that make up your or my body is sufficiently complicated, but it is simple compared to the chemistry of the reactions between the hundred million million molecules that make up a cell of average size.

The study of those reactions is the main task of biochemistry. Suffice it to say that we have begun to isolate many of the intermediate products of metabolism and the catalysts that govern the course of the reactions by which they are formed.

Chapter 9

Physiology of Development

The great difference between the two branches of physiology is this—that while adult physiology is dealing with stability, developmental physiology is dealing with change. In the former case, the general characters of the organism, in spite of alterations in detail, remain the same over long periods of time; and one of the most prominent features of the processes of adult life is that, as a whole, they constitute a self-regulating system, in which any excess in one direction automatically brings into play forces acting in the opposite direction.

But in developmental physiology, the organism is always passing from one phase into another, and even if some of the phases—like the tadpole stage of the frog—are of comparatively long duration, yet processes are always at work which at length will upset the self-regulative power of the system, and will make it change into something quite different. Besides these processes of normal development, however, others of the same general type may occur as the result of special circumstances.

The facts of regeneration, for instance, come under this head; and, as a matter of fact, some analysis of regeneration will be found to give the best starting-point for a discussion of developmental physiology in general. Regeneration is not a very familiar fact to most of us, since it is present to a very slight degree in ourselves and in the animals we know best.

But worms or polyps, or indeed most of the lower animals—if they could think—would find regeneration the natural, normal state of affairs, and the absence of regeneration a sad and abnormal failure. In a sense they would be right. Regeneration does at least seem to be a fundamental and original property of life, lost late and for special reasons. If you throw crumpled paper into a freshwater ditch, or set traps consisting of a bottle with boiled earth-worm for bait, you will probably succeed in catching large numbers of various kinds of freshwater Planarians—little leaf-shaped creatures of carnivorous habits and gliding movements.

Choose out the brown and black species, which are the hardiest. Transfer one to moist blotting-paper, and with a sharp knife cut it clean across. The experiment sounds cruel; but in reality you will have artificially encouraged the multiplication of the species. Each of the bits, replaced in water, will regenerate what is lost and so become transformed into a complete worm. The same would have happened if you had cut the animal into half a dozen cross-pieces instead of two, and equally whether you cut it down the middle or across.

If you persisted, you would discover two main facts in the course of your experiments—that any piece of any shape, provided it was above a certain quite small size, is capable of regenerating into a whole animal; and that this happens equally whether the regenerating process are fed or no. The same would have been true if instead of a planarian you had cut up the little polyp Hydra, and almost the same with the smaller Annelid worms such as Lumbriculus.

The same also holds for the microscopic single-celled Protozoa, although here the operations are much more difficult owing to the animals' small size. It is also true for the very early stages of development in many kinds of animal. In a newt either of the two first cells produced in segmentation may, if artificially separated, give rise to a whole embryo; the same holds in many animals for any one of the first four cells, or for most fragments of the late segmentation and blastula stages. Thus regeneration of the extreme type—what we may call complete regenerative capacity—is found in the simplest types of animal and the earliest stages of development.

It must therefore be thought of, not as a special property developed to meet the rare contingency of losing a limb or being bitten in two, but as something natural to living things, and present unless circumstances forbid it. Further facts confirm this view.

In animals of a rather higher grade of organization, such as the more complicated worms, crustacea, some insects, and some molluscs, the power to regenerate is still present, but it is limited. No longer can the animal be cut in half, or into a number of little bits, without losing its capacity for life; but so long as certain central and essential organs remain at work, very considerable losses, such as those of a limb or a tail, can be replaced.

This degree of regenerative power continues even into the land vertebrates. Although the frogs and toads do not, the lower or tailed amphibia still possess it. Even a mature newt or axolotl can grow not merely a new toe or new foot, but can repair the loss of a whole hind limb and limb-girdle. This we may call organ-regeneration. In such animals, the early stages in development of an organ such as a limb are capable of complete regeneration. Thus, by cutting a tadpole's limb-buds so that a number of separated bits are left, the animal may be made to grow more than the normal complement of legs. The same limitation of regeneration is seen, as might be expected, during development also.

For instance, while the amphibian in its segmentation and blastula stages has an almost complete regenerative capacity, the tadpole can only regenerate single organs, and the frog can do no more than heal wounds. Even in newts and salamanders regeneration is much more rapid in the tadpole than in the adult. Among reptilia, lizards are the only animals which possess even the power of organ-regeneration, and in them it is confined to the tail. When we look closely we find that here the original general power of regeneration inherent in living things has been retained in this organ for particular biological reasons, and has been combined with special adaptations to make it of greater value to the animal.

The lizard has the power of self-mutilation (or *autotomy* as it is often called). If you catch a common lizard by the tail you will find that after a moment or two of squirming, the tail, broken off short at the root, will be all that you have in your hand, and the rest of the lizard will be running off into safe hiding.

This is only made possible by a special structure of the vertebrae of the base of the tail; they have a crack on either side which penetrates almost to their centre, and the tailmuscles are so arranged that when one set of them contracts they pull at the further half of one of these vertebrae, and snap it off across the plane of weakness. The tail when broken off is, though doomed to death within a few hours, not yet dead; and it continues to execute the most violent wrigglings and jerkings for some time.

In Nature what occurs is apparently as follows.—When a bird or other enemy makes a pounce at the lizard, it will often only succeed in catching it by the tail. When this happens, the tail is broken off, and by its squirmings continues to keep the attention of the attacker while the lizard itself is making for safety.

The tail is lost and devoured, but the lizard grows a new one. If lizard-eating animals were rational they would probably confine themselves to tails, gathering the annual crop as we gather the annual crops of fruit off a fruit-tree.

In any case, the lizard often saves itself by sacrificing its tail (*how* often could be established by finding out how long tail-regeneration takes, and then by examining a number of lizards from one locality and noting how many were in process of growing new tails—an experiment well worth trying); and this desirable end has only become possible by adding the special mechanism for breaking off the tail to the primitive power for regeneration.

Crabs and many other Arthropods have this power of autotomy in their limbs, and some worms when handled break their whole body into fragments, each one of which will regenerate into a complete individual! One word about the power of regeneration in mammals and ourselves. Although we cannot grow new limbs, yet our wounds will heal, and this is still a real regenerative process; we may call it wound-healing or tissue-regeneration.

It is also to be remembered that in a certain sense the growth of such tissues as the epidermis is a continual regeneration—for something is continually being lost, and as continually being renewed. These last processes fall under the head of *normal* or *physiological regeneration.* A more striking example, perhaps, than the skin is to be seen in the antlers of deer—two masses of bone weighing several pounds which are shed and renewed each year. Two or three general principles emerge from these facts. The first—to recapitulate—is that regeneration is a primitive property of life. Most low organisms never stop growing, and the capacity for regeneration is clearly associated with the capacity for growth, since growth is necessary for many of the processes of regeneration.

Other changes that are at work in regeneration, however, have the character of rearrangements; for instance, in the formation of a whole newt out of one of the first two blastomeres of the newt's egg, the only process is one of regulation—of rearrangement of material which ought to have produced a half, so that it actually produces a whole of half size.

In the regeneration of a small piece of Planarian to form a whole worm, a certain amount of new growth takes place at the cut ends of the piece, but many of the changes involved are brought about by changes in the original piece. For this to happen, however, a certain degree of plasticity is needful—cells and tissues must be capable of altering their original character and adapting themselves to new conditions.

This also is largely a property of actively growing organs; further, it appears to be impossible if tissue-differentiation has reached too high a pitch.

Thus the gradual limiting of regenerative capacity which we find as a general rule both in the development of the individual and in the progress of life as a whole, is due partly to increased differentiation, partly to the fact that the higher animals, instead of growing indefinitely, show a sharply marked stoppage of growth on reaching the adult state. Regeneration is thus essentially the restoring of typical form, structure, and function; we might call it a special case of regulation.

The fact that a piece of Planarian without a mouth can yet regenerate head and tail is obviously interesting. The piece's capital can be readily transformed into cash for the undertaking of new enterprises—in other words, the living framework of its body can be broken down into foodmaterials which can then be used up in new growth elsewhere. In the higher animals, the living capital is locked up in non-negotiable forms to a much greater extent, although not so completely as is often imagined. This power of drawing upon the living tissues in case of need leads on to another remarkable power of Planarian and other low types.

If an intact animal of this sort is starved, it can continue to exist for periods of weeks and months by thus liquidating its capital for its day-by-day expenses. Since it draws on its own tissues for its daily energy-needs, it therefore must grow smaller and smaller. Planarians can continue this economical process until they have diminished their size to below that at which they hatched from the egg.

If they are given food again (at any stage in the proceedings) they will start growing once more, and recover their normal size and appearance. One very curious fact was early noted about Planarians thus reduced in size by starvation—namely, that their proportions gradually altered from those of the adult and became more like those of normal young worms.

This led to the idea that perhaps they not only *looked* young, but had really *become* young again—an idea which was easily put to experimental proof. Some species of Planarians reproduce mainly asexually by transverse division; after the worm has reached a certain length a new head is formed a little behind the centre of the body, and in front of this a split appears along which separation eventually takes place. A brood of worms from one such species was taken. When they had nearly reached full size, they were divided into two lots.

One lot, fed regularly, continued to grow and divide in normal fashion. The other lot was starved until the average size of its members was reduced to about half; then it was fed until it had recovered its original size; then starved again—and so on repeatedly.

The experiment was continued for a period of time in which the well-fed animals went through nineteen generations (a period which, though only occupying a few months for Planarians, would in human beings mean about six centuries—from Chaucer's time till to-day). During the whole of this period, none of the other lot reproduced at all, but all remained within the limits of size set for them by the experimenter, and, what is more, showed no signs whatever of age or of diminished vigour. In fact there is every reason to suppose that the experiment could have been continued indefinitely—in other words, that the individual Planarian, by proper treatment, can be made immortal.

The elixir of life was sought by the alchemists of medieval Europe for hundreds of years. These experiments show that it has at last been discovered; but, unfortunately, it is only effective with Planarians and some other animals, all of a low grade of organization.

Before we go further into the principles involved in this matter, we must mention that there is another way in which animals may surmount starvation and other unfavourable conditions. If an ordinary hydroid polyp like Obelia be kept in the laboratory, the hydranths, or separate organized individuals of the colony, will (unless the water is well aerated artificially) show a curious series of changes.

Their mouths will close and their tentacles become stumpier; eventually they will lose almost all signs of their previous differentiated structure and become converted into a rounded or egg-shaped bag with a few little knobs marking where the tentacles had once been.

They have lost differentiation—in other words, have undergone dedifferentiation. In Obelia the process is complicated by the fact that while differentiation is going on, the cells of which the hydranth is made up detach themselves from their neighbours and, leaving their fixed place in the tissues, migrate into the central cavity and down into the common stalk which connects all the hydranths. By this means the hydranth dwindles rapidly, and is reduced soon after dedifferentiation is complete, to a tiny knob in the bottom of its protective horny cup.

The animal has been resorbed by the stem. It is as if a house were to be unbuilt, by the bricks leaving their places in the walls and migrating into the garden. If, however, the hydranth is cut off at its base beforehand, the central cavity is soon filled up with detached cells, and after this, dedifferentiation alone occurs until the animal is reduced to a mere living bag tight packed with cells. These types of dedifferentiation are found in many low types of animals, such as various Coelenterates and Ascidians. Among higher forms partial dedifferentiation may be a normal mode of development.

For example, when the tadpole metamorphoses into the frog, some of its tissues start to dedifferentiate, for instance, those of the gills and tail. The process is complicated by the fact that phagocytosis, or the devouring of undesirable materials by white blood-cells, here reaches a far greater pitch of perfection than in lower animals, and, as a result, as soon as the cells have reached a certain degree of dedifferentiation, they are attacked and removed by the white blood-cells.

The disappearance of the tadpole's tail is thus due more to the activity of the phagocytes than to the migration of the tail-cells themselves. Similar methods of resorption of tissues are found when, for instance, a tumour or a graft disappears in a human being. To return to regeneration, and to another of its aspects, it is found that the presence of one part is often necessary for the regeneration of another part.

In Planarians, for instance, the mouth opens at the end of a sort of trunk, the muscular pharynx, which can be protruded from the centre of the underside. If a small piece of the flatworm is cut out by two transverse cuts, it depends on a number of circumstances whether a head shall be formed or not; for instance, a low temperature near freezing, or exposure to dilute alcohol or other poisons (the experiment is easy to carry out) will reduce the frequency of head-formation, or even suppress it altogether. If a head is produced, such a piece will always proceed to the formation of a pharynx. But if no head appears, the formation of a pharynx depends on the original position of the piece in the body.

If it was cut from in front of the original pharynx, it will often regenerate a pharynx, even in the absence of a head; but if it came from behind the original pharynx, a new pharynx will never be formed unless a head forms first. In other words, a new pharynx will be regenerated in the centre of a piece of Planarian provided that there exists at the front end of the piece a region which is normally anterior to a pharynx—whether this anterior region be an old part of the body or a new regenerated head. Thus, the head or front region of the body exerts some remarkable influence upon a distant region of the rest of the piece, making it construct an organ that it could not otherwise produce. It makes things make themselves. Exactly how this effect is produced we do not know.

Possibly it is through the agency of the nervous system. In any event, it is clearly a property of very great importance, and also one which seems to be of quite general occurrence. For instance, when the eye-stalk of a prawn is cut off, a new one like the old is usually regenerated. But when, in addition to the eye-stalk being amputated, the optic ganglion, or part of the brain to which run the nerves from the eye, is also cut out, the organ which is regenerated bears no resemblance to an eye and eye-stalk, but has precisely the form of part or all of the first antenna or feeler, the appendage which normally comes next behind the eye. One other fact gained from the study of regeneration needs to be considered before the problems of normal development.

If one of a pair of organs be removed, remarkable results often follow. For instance, in certain diseases, one kidney may have to be cut out. When this is done, the other begins to grow, and at the end of a few months weighs almost as much as the original pair together. This enlargement follows as a direct result of the increased demands made upon the functions of the surviving kidney, and is therefore known as compensatory functional overgrowth.

There is no regeneration of the missing organ, but a compensatory growth of its mate, resulting, one may say, in a *regeneration of function*. This functional overgrowth may be found not only to compensate for the loss of one member of a pair of organs, but also when part of a single organ is removed, or when an extra demand is made upon an organ without removal of anything.

If, for instance, most of the liver or pancreas or reproductive organs are removed, the bits that are left will start to grow rapidly, until they are large enough to cope with the demands of the body. Or, as illustrative of the second case, the normal thyroid secretion is very rich in iodine.

When the amount of iodine in the food and drink falls below a certain quantity, the demands of the body for more iodine-containing secretion react on the thyroid; this begins to grow, producing more total amount of secretion as an attempt at compensation for its poverty in iodine. It may grow to a relatively huge size, producing a great swelling, called a goitre, in the neck. If a little iodine is given to a patient with such a goitre, the demand ceases, the reverse process sets in, and the swelling is reduced.

A similar process takes place in muscles when extra demands are made upon them. Everybody knows how heavy exercise 'develops the muscles', and, after a preliminary period of loss of weight due to utilization of the reserves of fat, makes a man put on flesh. The actual size of the muscles increases as the result of exercise.

With this information we can now attempt some account of the normal development. It is found that the early development of a vertebrate such as a frog or newt falls, from our present standpoint, into three main periods. The first is essentially a period of the division and rearrangement of already existing materials, and takes the fertilized egg through segmentation and some way into germ-layer formation.

The second is mainly one of differentiation—visible structure appears, the organs are blocked out, and tissues become different each from the other. Finally, in the third, the two most important features are growth, and the welding of the differentiated parts into a working whole—by mutual influence of organs upon each other and by an adaptation of separate organs to the demands upon them through the effects of function upon growth.

This last phase may be taken first, since it is in many ways the most familiar. Think of a young organism—a recently hatched tadpole, a puppy, a child. Although it has already a characteristic organization, yet this organization is still plastic within wide limits.

The final shape and size of its bones, its muscles, its sinews and tendons, its glands, its digestive tube, even of parts of its central nervous system, depend very largely upon the amount and the kind of use to which they are put during development. If one of a puppy's fore-limbs is tied up soon after birth so that it is never used, the dog will manage to get about very well with only three legs. But the leg which has not been used will be very different from the others.

The bones of its skeleton will grow almost to the normal length, but will never attain more than about half their normal thickness, nor will the arrangement of bony struts and stays at either end of the shaft (which in an ordinary bone are developed so as to meet both vertical and sideways strain in the mechanically most perfect way) be properly formed. The muscles of the leg, too, and their tendons will be very small. Not only this, but the nerve-cells of that region of the brain which controls the movements of the limb will be undersized. The power of bone to respond to the strains to which it is exposed, is of practical value.

If, for instance, one of the small bones or phalanges of a finger is crushed and has to be removed, a small piece of bone from some other region can (if it include some of the periost layer from which new bone-cells are produced) be grafted in to take its place.

After a year or two the irregular splinter will have come to bear a marked, though not complete, resemblance to the bone for which it is doing duty. One of the most striking adaptations in the body is the fact that every tendon which attaches a muscle to a bone not only is of the proper size for the strain which the muscle can exert upon it, but is composed of parallel fibres which run precisely in the region of greatest stress.

How can this beautiful relation of mechanism to function be supposed to have originated by natural causes? Does not the purposeful adaptation, different in detail for each separate tendon, demand the intervention of some supernatural power? Research, however, has in great measure removed such difficulties; for it has shown that all the separate detailed adaptations are the direct consequence of one inherent property of the tissue of which tendons are made. If a piece is cut out from a long tendon, such as the Achilles tendon which passes from the muscle of the calf over the heel to be fastened to the sole of the foot, it will after a time regénerate.

Connective tissue grows out into the wound, the cells elongate and form fibres in the direction of the tendon, make attachment to the cut ends, and reproduce a structure essentially like that which was originally present. But if by one means or another the calf-muscle is put out of action (as when its nerve has been severed) then, although the gap will be filled by a new growth of connective tissue, the fibres will not become arranged in any particular direction, and no tendon-like structure will be formed.

Further, when in such an animal (the main experiments were carried out in rabbits) a piece of clean silk was healed across the middle of the wound, at right angles to the original tendon, and then, by winding it up very slowly on a screw, increasing tension was put on one end of it, it was found that fibres were formed in the new connective tissue, but parallel to the silk, and therefore useless for any restoration of normal function. From these facts (supported by many others) it is clear that tendon-fibres are formed from connective tissue when the connective tissue is frequently stretched, and further, that they are formed parallel to the stretching force.

This stretching force is usually the force of muscular contraction (in growing animals there will also be found stretching due to growth of bones); so that tendons will be formed from ordinary connective tissue when muscles begin to contract, and will be 'adapted' to the direction and amount of stretching exerted by the muscle. In any case, all the apparent design seen in the adaptation of each several tendon to the demands made upon it, is apparently a direct result of this one reaction of connective tissue to tension.

To sum up, we may say that certainly most of the tissues of the body—possibly all—respond either throughout life or at least during growth (while their cells are still capable of multiplication) to the demands made upon them, and respond on the whole very advantageously from the point of view of the whole animal.

The skeleton, the muscles, tendons, many and probably all glands, the central nervous system—all these we have already seen to show functional adaptation. So does the skin. Everybody knows that (after the blister stage has been negotiated) continual rubbing will make the skin thicken. 'Horny-handed' is so hackneyed an epithet for manual labourers as to have almost become mere journalese, and every oarsman demonstrates the fact in his own person. So, too, the circulatory system responds to functional demands.

If Everest is ever climbed, it will probably be because men can become 'acclimatized' to high altitudes and consequent low oxygen supply, largely, it would seem, through an extra production of red bloodcorpuscles by the bone-marrow and in consequence a greater capacity of the blood for capturing and transporting what oxygen there is. Furthermore, the size of a blood-vessel, both diameter of cavity and thickness of wall, is determined very largely-perhaps wholly—by the amount of blood which is pumped through it, and the pressure at which it is pumped.

Cases are known in which large arteries become blocked up. If the blocking-up process is not too rapid, one or more small branches of the vessel will gradually enlarge, and produce channels capable of transporting the requisite amount of blood without difficulty.

This has been known to happen even when it is the human aorta, the main and indeed the only large artery springing from the arterial side of the heart, which has been blocked. The same kind of adaptation is also true of the protective reactions of immunity.

It is a familiar enough fact that children who have had measles once do not generally catch it again, but this is only one among innumerable instances, such as the over-production of diphtheria antitoxin by horses injected with small doses of the toxin or poison produced by diphtheria germs, which has reduced the disease from a cruel scourge to a serious unpleasantness, or the benefits given by vaccination.

All these again depend on one single property of higher animals; when unaccustomed proteins enter their system, they can (if the foreign substances are not in too great amount, and provided certain other conditions are fulfilled) not only destroy or inactivate them, but in the process generate an excess of the destroying substance or *antibody*, which remains for a longer or shorter time so that a second invasion can be more rapidly and effectively crushed.

Although many of these adaptations are possible throughout life, others can only occur during development, and most of them are essential for normal development, since only through them do the size and structure of the various organs come to correspond to their function and to the physical and chemical stresses imposed on them.

There is, however, another type of mechanism which is active through the later periods of development and plays an important role not only in this functional adaptation but also in fusing the parts of the organism into a more unified whole, and that is the endocrine system or sum total of the ductless glands.

Many of them, such as the thyroid and the pituitary, have indispensable functions as regards proper growth. The thyroid regulates the rate of metabolism, increased amount of thyroid secretion producing an increase of oxidation in the body. As might be expected, this function seems to be of more importance to the well-being of animals such as mammals, with a normal high metabolism (great heatproduction and constant high temperature), than to those like fish or frogs, which do not have to be constantly generating large quantities of heat to keep their temperature above that of their surroundings.

Within the mammals, the absence or failure of the thyroid makes more difference to an adult man than to an adult sheep: the thyroidless sheep is but slightly abnormal, while the thyroidless man becomes notably fat, sluggish, and slow-minded.

But its absence in the child, in whom growth and activity should be much greater, makes far more difference again than in the grown-up man; the thyroidless human child or *crétin* usually dies early, can only grow into a stunted dwarf, and never develops normal intelligence, the brain of man apparently requiring a high level of metabolism for its unfolding.

In Amphibia, metamorphosis from larva (tadpole) to adult is under the control of the thyroid. Tadpoles with their thyroid cut out refuse to be transformed into frogs, and grow to giant size in the water, while if a dose of thyroid is given to quite young tadpoles, they will metamorphose into froglets no larger than flies. The changes which take place in body and mind at puberty are also under the control of the endocrine system. The immediate control is exerted by the gonad itself.

It is important to find that a whole set of separate changes can be brought about by a single change in the ductless glands, and further, that all these changes are in the long run related to one function, that of reproduction. That a single original alteration in the chemical composition of the blood, an alteration appearing inevitably at the end of a definite span of growth, should bring about an increased growth-rate, the formation of mature spermatozoa in the testes, the appearance of hair on the face and body, and a marked change in mental outlook and temperament, is one of the most striking facts in biology.

All the ductless glands have this in common, that their secretions are carried all over the body by the blood, so that they can exercise a number of different effects simultaneously upon the most diverse organs. It is fairly obvious, however, that for all the importance of function and of the endocrine system in development, their task is essentially one of regulating the degree of activity of processes already in being, or of directing them along particular channels. Bones exist in the absence of a pituitary or of pressure upon them, and muscles without being used.

It is only a small part of the adult man's growth of hair that is dependent upon the testis; metabolism continues in the absence of a thyroid, and the first development of a gland or blood-vessel takes place before its use begins.

Another point that should not be forgotten is that in the long run the differences in shape and size between different animals depend almost entirely upon differences in the amount and direction of growth in different regions during development. In very early life the hind-limbs of a whale grow much less fast, compared with the body, than do those of a man, and those of a man much less fast than those of a kangaroo.

Some idea of the changes which would be produced in human beings if the proportions between general growth in length and growth in breadth were rather different from what they actually are, can be obtained by visiting the distorting mirrors at a fair or circus.

Or one organ may be affected differently from the rest; the effect of this on general proportions may be seen by looking at oneself in the broadening mirror, first with the arms held against the sides, and then outspread at right angles to the body.

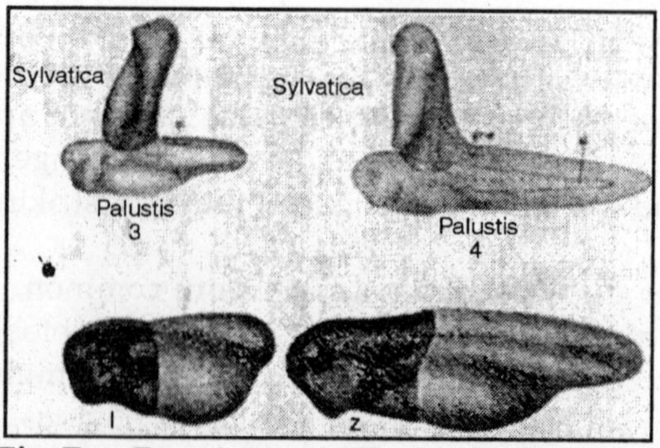

Fig. Two Experiments in Grafting in Tadpoles.

- The front half of an embryo of one American species of frog (*Rana sylvatica*) has been cut off and grafted on to the hind half of another, lighter-coloured species (*R. palustris*).
- The compound animal (chimaera) grows quite normally. The lateral line, which originates near the head, has grown down from the *sylvatica* component on to the *palustris* trunk. Compound individuals like this have been reared through metamorphosis. 3 and 4. The front part of a *sylvatica* embryo has been grafted on to the back of a *palustris* embryo: this combination also has continued development. The *sylvatica* lateral line (below *si*), on reaching the *palustris* component, has bent round into correct position.

If in the embryo of the fish Argyropelecus the tendency to growth in length had been a little less, and the main direction of dorso-ventral growth were not quite at right angles to the main direction of growth in length, it would have developed into a passable imitation of another kind of fish. No doubt such slight alterations in growth are at the bottom of much evolutionary change.

How do these more fundamental structures and functions originate to provide the primal basis to be shaped by the secondary moulding forces we have described? Although much is still not clear, yet the main outlines of the answer emerge when we look at the next previous stage of development, the stage of differentiation. This starts during gastrulation, and ends as the various organs become capable of exercising their functions.

Its great characteristic is the sudden appearance, in the hitherto comparatively simple germ, of a number of chemically different regions, each one of which is irrevocably destined to generate one particular kind of organ.

What is more, this will usually be accomplished whether the developing organ-rudiment is in its proper relation to the rest of the organism or not. If the late gastrula is cut in two so as to separate front from hind region, each part (after healing the wound) will proceed to develop just those organs which it would have done if the embryo had been left intact. We are presented with an anterior half-embryo without any of the hinder organs, and a posterior half-embryo with tail, hind limb-buds, and most of the trunk, but no head-regions, gills, or heart.

There is, in fact, in this stage *no* regeneration whatever of missing parts; yet, in spite of this, both halves appear quite healthy, and only begin to degenerate at a stage when the organs should be beginning to work and the animal to fend for itself. Or a piece of one embryo may be grafted on to another embryo, even of a different species, and both will continue to develop. One actual 'chimaera' has been artificially produced consisting of the front half of one kind of frog and the posterior half of another, which grew and metamorphosed quite happily. Or if a small part of the future brain-region be cut out just after the gastrula stage, and grafted in again in reversed position, it will produce what it would have produced in an untouched embryo, although now of course the organs to which it gives rise will be misplaced.

The eye, for instance, which is formed by an outgrowth from the brain, may by this means be made to develop behind the ear, and yet have normal structure. This will happen in spite of the fact that no visible trace of even the earliest eye-rudiment was present when the operation was performed. Before gastrulation, on the other hand, no such irreversible changes, inevitably determining the future appearance of organs, have yet set in. Cutting the developing egg in half in the earliest stages has a totally different result from cutting it in half after gastrulation has set in.

In the former case, as we have seen, one or both halves may become reorganized into a whole, and produce an embryo normal in every respect save size. One very remarkable fact has recently been discovered, namely, that the sudden change that takes place in a few hours during gastrulation in an animal like a frog or newt, during which the germ loses its capacity for regulation and becomes a sort of jig-saw of separate parts, each predetermined to produce just one particular organ of the future tadpole, occurs under the influence of one small region of the embryo.

This is the region just above the dorsal lip of the blastopore. By ingenious microscopic operations, even tiny bits of a newt's egg can be removed and grafted into other places on the same or other embryos. If this dorsal-lip region be taken and grafted into the flank region of another developing egg, it is found to cause the development of an extra set of organs, in addition to the one produced in relation to the host egg's own dorsal lip. An extra nerve-tube, notochord, muscle segments, kidney tubes, and gut may be thus induced to form—in other words, the beginnings of a whole second embryo.

What is more, they are not formed by the grafted piece itself, but in the host-tissues near it. Again, we are reminded of Mother Carey in Kingsley *Water Babies,* who 'made things make themselves'. No other region has this power, and so the title of 'organizer' has been bestowed upon the dorsal-lip zone.

It is interesting to recall that a similar sort of power was exerted by the head of a Planarian in inducing the formation of a mouth and pharynx.

Another fact is that the dorsal-lip region at this period is the most actively working part of the embryo, with very rapid cell-division. It seems probable that this great activity has something to do with its organizing capacity. Other interesting results, also apparently concerned with differences of activity in different parts of the developing egg, have been obtained with still earlier stages.

Although the period up till early gastrulation is of course an essential link in the developmental chain, no important steps in the construction of the details of the animal's typical organization are being taken so long as it lasts. The original comparatively simple organization of the zygote is not much altered; and the most essential change is the cutting up of the huge and unwieldy single cell, the fertilized ovum, into hundreds of small cells.

In the first place, these are far more mechanically convenient, just as a hundredweight of bricks is suitable building material, while a single giant brick weighing a hundredweight would be useless.

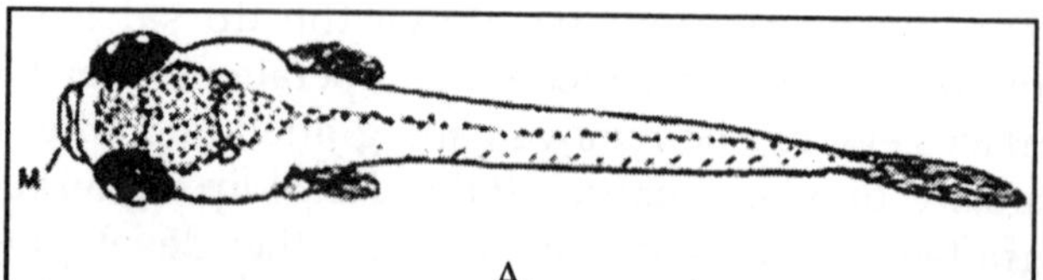

A.

Normal Larva, with Anteriorly Placed Mouth (M).

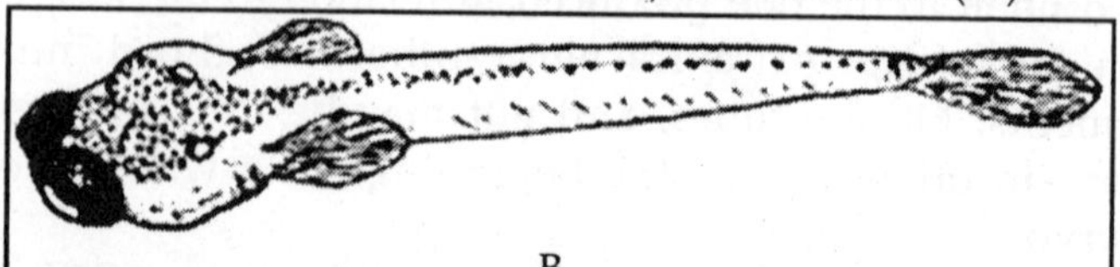

B.

Incompletely Cyclopean Larva, with the two Eyes Joined and Occupying the Position Usually taken by the Mouth.

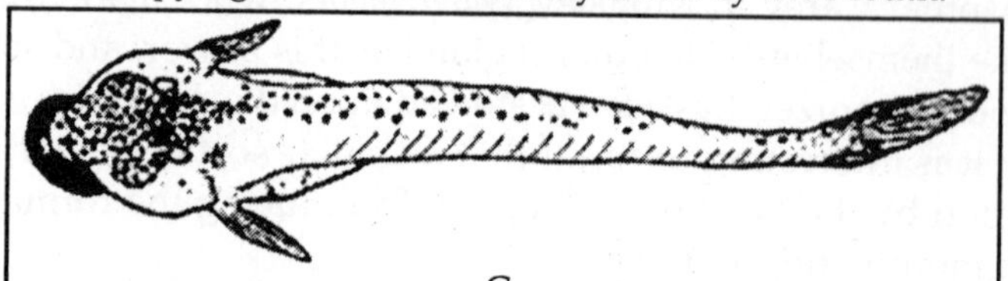

C.

Completely Cyclopean Larva, with Single Antero-median Eye. Dorsal aspect.

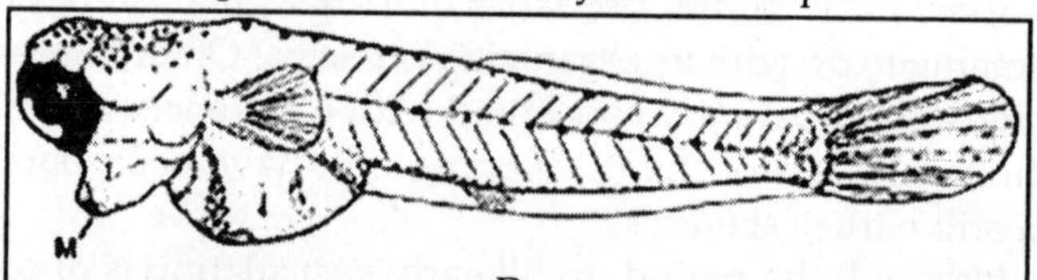

D.

Lateral aspect of same, Showing the Ventral Mouth (M).

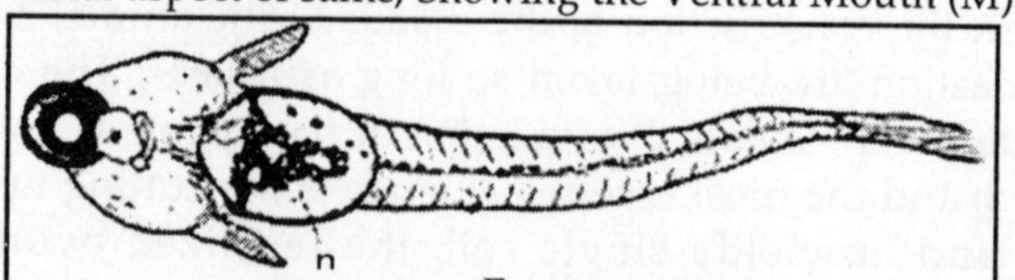

E.

Ventral aspect of same; Ys, Yolk Sac. (From Stockard).

Fig. Free-swimming Larvae of the Top-minnow.

And then each is a relatively independent unit, with its own cell-membrane as frontier, so that it is much easier to localize a number of separate chemical processes in different parts of an embryo divided up into cells than it would be to localize them if the embryo were all one large cell, in the same sort of way that the possession of a number of small pots and pans makes easier the task of preparing an elaborate dinner.

To choose another parallel from human affairs for the second and third phases of development, we might say that the second corresponds to the discovery and invention of new processes, new machines, new products, while the third corresponds to the growth of the industries based on these discoveries, a growth which is regulated by the laws of supply and demand, by the customs and laws of the country, and by the state of world affairs, but can only take place once the new inventions have been made.

In conclusion, it is worth mentioning that every cell of the body appears to receive the same complement of chromosomes, the original set possessed by the fertilized egg being exactly duplicated at each cell division, so that each cell has a 'copy' of it.

This at first sight seems to make differentiation hard to understand; but perhaps a musical analogy will make it clearer. The performance of a piece of pianoforte music demands a pianoforte. But we might have a hundred identical pianos in a row, from each of which would be emanating a wholly different piece of music: that would depend upon the players. For the pieces of music substitute the different organs of the embryo; then the pianos are the chromosome-sets and the players are represented by the different conditions in different regions.

For instance, at the close of segmentation some cells are small and poor in yolk, those at the opposite end are large and yolky, some are at the surface, others tucked away in the interior; and the differences between different parts are increased after gastrulation.

In animals progressive development generally comes to a close with the attainment of the adult condition. In man, however, this need not be so in respect of one character. Mental development may continue; in other words, the contents of the mind may increase and its organization become more complex and more perfect, so long as maturity continues and senility has not set in.

Of course this is by no means always very marked, and equally of course something of the same kind is seen in the higher vertebrates which can profit by experience. But the extent of the possible development open to man is none the less very striking.

One of the chief biological differences between man and all other organisms is that, through his power of thinking and learning, his development and differentiation can continue through the whole of life.

Chapter 10

Evolution of Physical Structure

Birds and mammals both possess a constant temperature which is ordinarily well above that of their surroundings. They should, therefore, be equipped with an arrangement for preventing too rapid loss of heat; and this is provided by the feathers in one case, the hairs in the other. Swimming birds, almost without exception, are webfooted or lobe-footed.

The Jacana, which seeks its food on the floating leaves of water-plants, has enormous toes which distribute its weight as do skis or snow-shoes on the thin crust of snow. Indeed, in general the feet of birds are remarkably well adapted to the life which their owners lead.

The egg-eating snake, Dasypeltis, possesses a special mechanism by which it can temporarily dislocate its jaws to swallow an egg whole.

In addition, one of its neck vertebrae bears a downwardly directed spine which protrudes into the gullet, and is used to crack the egg where none of its contents will be wasted. The larvae of crabs live near the surface of the sea, and are provided with long spines which increase friction and make them less ready to sink.

But the most striking examples are those known as convergence, where a similar mode of life produces similar effects on quite unrelated animals.

For instance, it is advantageous for many animals to be invisible against their surroundings, either to escape their enemies or to approach their prey unobserved. And we find that in polar latitudes many animals are white, at any rate in winter, in deserts many are sandy, in undergrowth many are blotched and streaked so as to break up their outline, and harmonize with the tangle.

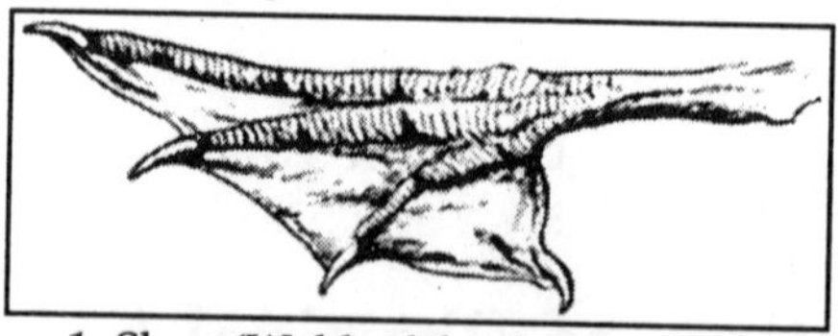

1. Shag (Webbed for Swimming);

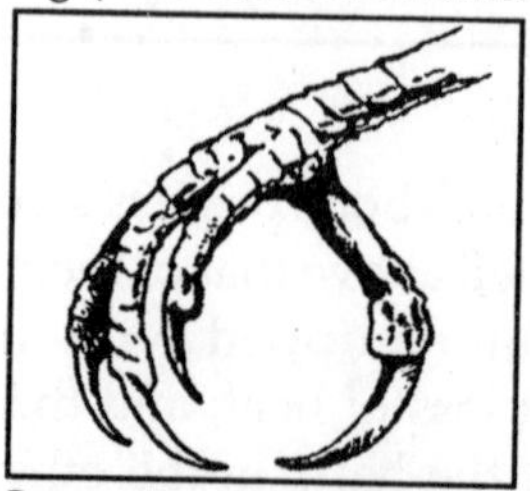

2. Crow (Perching, Grasping);

3. Ptarmigan (Running; 'Stocking' of Feathers for Warmth);

4. Wild Jungle Fowl (Walking, Scratching; Spur for Fighting);

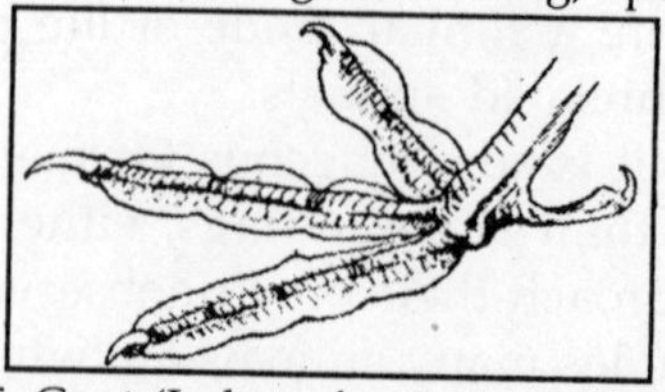

5. Coot (Lobate for Swimming);

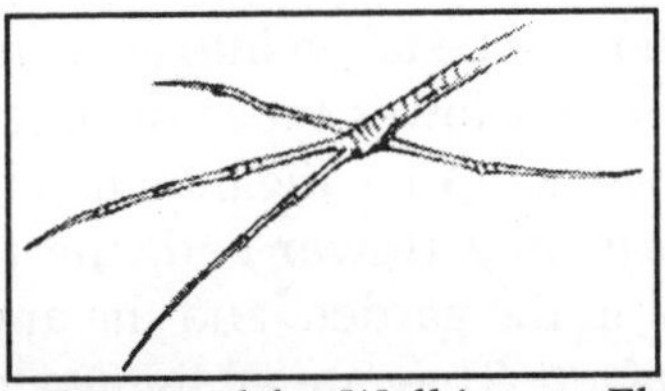

6. Jacana (Toes Elongated for Walking on Floating Leaves);

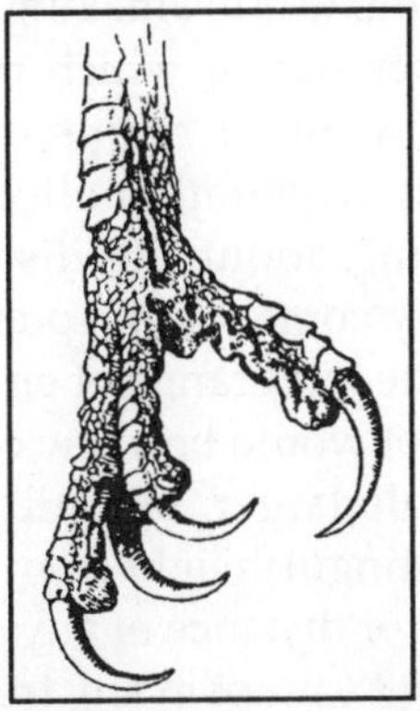

7. Sea-eagle (Raptorial for Killing and Holding Prey).

Fig. Adaptation of Structure to Mode of Life as Illustrated By The Feet of Various Birds:

Others, again, escape their enemies by mimicry, as it is called—by resembling other animals which are dangerous or distasteful. Thus wasps, which advertise their sting by their bright black and yellow pattern, are imitated (of course quite unconsciously) by a large number of different sorts of perfectly harmless insects both as regards pattern and bodyform. Ants may be mimicked by many other insects.

Another type of convergence concerns shape. Every one is familiar with the typical fish shape—the stream lines of the body, permitting of rapid motion in water. When other vertebrates have taken to the sea, they too have evolved into a similar shape, as is seen in the Ichthyosaurs, a group of reptiles, and the whales and porpoises, a group of mammals. This adaptation of every part of an organism to its role, of every whole organism to its mode of life, is universal.

It is indeed the direct and most obvious outcome of Natural Selection. But there is another fact, of perhaps greater importance, which must be taken into account, and that is the existence of higher and lower types of organisms. Within a few square yards we may have the man in a house, the dog in the yard, the worm in a flower-bed, the fish (probably a goldfish) in a pool in the garden, and the amoeba also in the pool, or in a water-butt. They are all living close together, but their effective environments are amazingly different in extent.

Most of the happenings to which the amoeba responds accurately take place within a radius of a millimetre or so; when stimuli like light or vibration affect it, it has no means of discovering anything about the distance or the kind of source from which they come, but responds simply to light or to vibration as such. The total range of environment which the Hydra could conceive of would be a few centimeters each way.

The worm is a little larger, but still without any special sense-organs; it can distinguish light from darkness, but cannot see the shape or colour or distance of anything; it can tell when the earth is vibrating, but cannot in any true sense *hear*, because it cannot distinguish tones; it cannot begin to control the environment in the same sort of way as man controls it, because it is not even in contact with most of that environment, locked away from the happenings of the outer world in a windowless existence which is almost incomprehensible to us who are provided with efficient sense-organs.

The fish can see images, but its vision of anything outside the water is very limited owing to the water surface, which if it is rough prevents vision across it. In addition, it is of course confined to water, and so, in this case, to a little pool of a few feet radius. The dog can see, can hear and smell very well, and can roam over the surface of the earth. Its environment as perceived by its senses is as extensive as that perceived by man's senses; but it cannot understand it in the same way. For instance, we can be pretty sure that no dog could ever come to understand the difference in distance between a cloud, the moon, the sun, and the stars.

In addition, its brain cannot make the same reasoned relations between its sense-impressions as can man. Although it can and does learn, it can only learn in an unintelligent way, making associations as they come. It does not appear capable of thinking in abstract terms of cause and effect. Its environment must seem both more limited and more chaotic than the man's. The man, if he takes the trouble to be interested in the environment in which he lives, finds it a very marvellous one. He can get information, by means of a microscope, of invisible things under his nose; he can also obtain information, by spectroscope and telescope and mathematical calculation, about the composition, size, and speed of stars hundreds of light-years away.

He can know by letter and newspaper what others are thinking and doing, through history he can enlarge his past far beyond the limits of his single lifetime, and he can make prophecies, sometimes (like those of eclipses and comets) of perfect accuracy, concerning the future. Also he can relate the different parts of his environment to each other and to general principles. His environment is enormously greater, both in space and time, than the dog's—let alone the amoeba's; and it is much more intelligible. In addition, the six organisms are of very different sizes and degrees of complication. Amoeba consists of one cell, hydra several thousand, the worm many hundreds of thousands, a man millions of millions.

There are many more kinds of cells in man, dog, or fish, than in hydra or even worm. As we ascend the series, we find even greater independence of external forces. Amoeba and hydra are at the mercy of floods, currents, droughts. A worm is limited to a very small section of the soil. A fish is wide-ranging, and can change its abode at the onset of unfavourable circumstances, but is confined to water. The dog can range on land, and, finally, man is at home in every latitude, and has mastered sea and air as well as earth. When we look carefully into the matter, we can see that here, too, the differences between members of the series can be thought of in relation to the environment, but in a much more general way than is the case with special adaptations.

Using the word environment to denote the whole series of events and processes with which life can come into contact, and not merely the particular environment of one particular organism, we can say that some animals have more control over the environment than others, and are more independent of it. The savage is more at the mercy of the elements than is civilized man; civilization is learning to control floods and droughts, to make a pathway of communication of the same sea which to the savage is an impassable barrier. The dog in its turn is not able to cultivate the ground or to kill such varied game as the savage.

The fish can exert far fewer different movements than the dog, and is much less able to profit by experience. The earthworm is not only without specialized organs of locomotion, but also lacks all organs of special sense. So far as independence goes, it is well to remember that amoeba and hydra are bathed, over the whole of their absorptive surface, by the water in which they live, and that accordingly any changes in the composition of that water act immediately upon the vital processes of the animals. In addition, they do not possess any mechanism for regulating their temperature, and so must live slow or fast according as their surroundings are cold or hot.

In man or any mammal or bird, both the temperature and the chemical composition of the fluid which is in contact with all the cells of the body are, as we have seen, kept constant with an extraordinary degree of accuracy, and special devices exist for preventing changes in the outer world from exerting their full effect upon the vital processes of the body. In brief, we may say that high and low organisms can be distinguished by the degree of their control over and their independence of environment.

This difference in independence and control is reflected in their structure and their capacity for self-regulation and, in the mental sphere, by the degree of knowledge of the outer world which their sense-organs and brains permit, and in all probability by the intensity of their emotions. From what we know about evolution it is clear that the highest organisms have developed latest in time. This we can actually see happening as we trace back the history of life in the fossils; it is a probability which amounts for all practical purposes to certainty that the converse is true, and that although the early stages of the earth's history as written in the rocks and fossils are now undecipherable, yet that the first forms in which life appeared were low organisms. At the beginning, then, there were only low organisms, today there exist all gradations between the highest organisms and the lowest; as we shall see later, undoubtedly many organisms have degenerated during evolution from a higher to a lower condition.

If we look back into the history of fossils, and investigate what forms of life were present before and after the development of some new high type, such as man or the mammals, we shall almost always find that the new type has simply been added to the previously existing types. For instance, the reptiles were the dominant land animals in the secondary period, before the development of the higher or placental mammals; when these were evolved, just before the beginning of the tertiary period, although they speedily became dominant on land, and although many species and even whole sub-groups of reptiles were extinguished, yet the reptilian type as a whole did not perish from off the face of the earth, but continued to exist as well as the mammals.

In the same way, although the advent of man sealed the death-warrant of a great many species of other mammals, reduced the total number of lower mammals very considerably, and deposed them from their previous dominant position, yet lower mammals still exist in abundance, and will undoubtedly continue to do so. Thus we cannot say that evolution consists simply in the development of higher from lower forms of life; it consists in raising the upper level of organization reached by living matter, while still permitting the lower types of organization to survive. This general direction to be found in evolution, this gradual rise in the upper level of control and independence to be observed in living things with the passage of time, may be called *evolutionary* or *biological progress*.

It is obviously of great importance, and can be seen, on reflection, to be another necessary consequence of the struggle for existence. This improvement has been brought about in two main ways, which we may call *aggregation* and *individuation*. Individuation is the improvement of the separate unit, as seen, for example, in the series Hydra—Earth-worm—Frog—Man. Aggregation is the joining together of a number of separate units to form a super-unit, as when coral polyps unite to form a colony.

This is often followed by division of labour among the various units, which of course is the beginning of individuation for the super-unit, the turning of a mere aggregate into an individual.

Let us take as an obvious example of biological progress the colonization of the land by vertebrates. As a matter of verifiable fact, the sea was already peopled with highly developed fish before the first amphibians appeared on land. In other words, while there existed great competition among vertebrates in the sea, this competition did not as yet exist on the land. Clearly, then, it would be a biological advantage to any species if it were to vary in such a way as to make it able to live on land, for then unchecked multiplication would be possible for it, and it would have fewer enemies.

Variations in this direction would thus tend to be preserved; in other words, this particular step in biological progress would be favoured. As a matter of fact the step is a very large one, and progress was inevitably slow. The Amphibia did not arrive at a complete solution of the problem of leading a terrestrial existence. Their skin is moist, they are usually confined to wet places even in their adult existence, and the earlier part of their life is almost invariably spent actually in water, in the shape of a tadpole larva.

With the evolution of the Amphibia the fringe of the dry land, the territory between land and water, had been conquered, but not the dry land as a whole. Once again, after millions of years during which the Amphibia were the highest vertebrate type, evolving life was confronted with a situation in which a premium was placed upon a further advance: animals born with heritable variations making it possible for them to live farther and more permanently away from water would become heirs of a rich unoccupied territory.

Thus it was that the Amphibia, after themselves arising from the fish, in their turn gave origin to the reptiles. In the same way, in the continual struggle that is going on in mammals or birds between herbivore and carnivore, pursuer and pursued, each new advance in speed and size or strength in one party to the conflict, must call forth a corresponding advance in the other, if it is not to go under in the struggle and become extinct.

The striking improvement in the running powers of the horse family during its evolution, evinced in increase of size, lengthening of the legs, reduction in the number of the digits, and development of a well-formed hoof, and the similar improvements that occurred in other Ungulates, were accompanied by a corresponding increase in the size, speed, and power of the group of Carnivora during the same geological periods.

Each was at the same time the cause and the effect of the other. A precisely similar state of affairs is often to be seen in the evolution of the tools and weapons and machines of man. For instance, in naval history, the increase throughout the nineteenth century of the range and piercing power of projectiles on the one hand, of the thickness and resistance of armour-plate on the other, provides a very exact parallel with the simultaneous increase of speed and strength in both carnivores and their prey.

In Nelson's time, the men-of-war were built of unsheathed wood, and the guns fired round iron balls, with a maximum range of a few hundred yards.

To-day, battleships carry 15-inch guns which fling steelcapped and pointed projectiles, laden with high explosive, for a dozen miles or so, while the armour-plating of heavily protected ships may now reach a thickness of a foot or even more of specially-treated nickel steel.

The earliest types are placed below. The number of digits on both limbs is slowly reduced; the middle digit is enlarged and specialized as a hoof; the fore-arm is strengthened by the fusion of radius and ulna, the ulna in its distal part disappearing; the hind limb is strengthened by the disappearance of the fibula and corresponding enlargement of the tibia; the length of the tooth is increased; its grinding surface becomes more complex. In addition there is an increase of size and change of proportions.

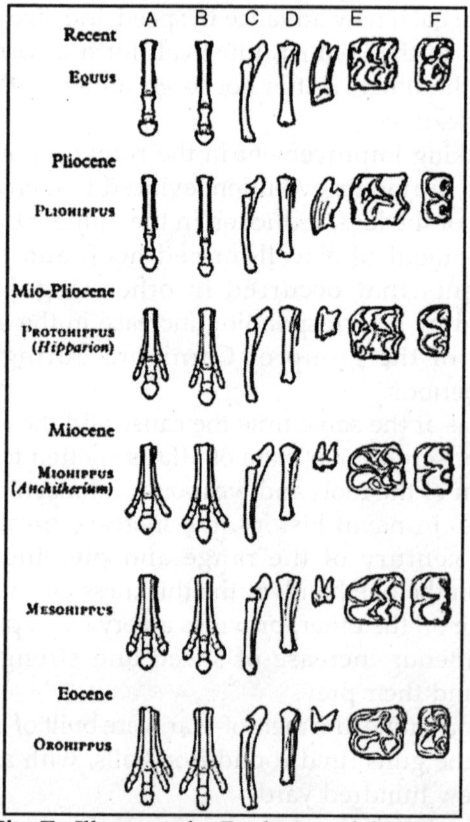

Fig. To Illustrate the Evolution of the Horse, From the Eocene Period Till To-day.

During the interval, progress has been steady and gradual in both departments of naval warfare, each advance in efficiency of guns being the stimulus for new invention in the methods of protection, and vice versa. It is when there is a general increase of the animal's powers of control that we speak of progress; when the increase is in one special particular only, we speak of specialization. For instance, the horse is specialized for running, the mole for burrowing, the bat for flying, the whale for a marine life, the lion for catching and devouring large animals, the sloth for living in trees.

Each of these animals is well adapted for its particular mode of life, but each is by that very adaptation quite cut off from leading the life of any other. During the late Secondary period there existed similarly specialized types of animals. For example, the Ostrich Dinosaur was adapted for running, the Ichthyosaur for life in the sea, the Tyrannosaur for preying on large animals, the Pterodactyl for flight, and so on.

But all the former list of animals are mammals, all the latter were reptiles. The mammals are higher in that they possess proper temperature-regulation, for instance, and better prenatal care of their young, as well as in many other points.

Thus, while the types of specialization, or of adaptation to particular modes of life, are somewhat similar in the two cases, yet each member of the first group is higher than any member of the second, because its general organization is on a more efficient level.

Whenever a new group of animals is evolved, it is found that its members soon become specialized in different directions, thus filling up different vacant places in the economy of Nature. This adaptation to different modes of life, while, as we have seen, we call it *specialization* when we are thinking only of one species of animal, is called *adaptive radiation* when we are thinking of the group as a whole.

All fish, for instance, breathe by gills, possess fins as limbs, and have other characteristics in common, so that any member of the group can be at once recognized by a brief examination of its structure. Yet the detailed form of different species of fish is extremely varied.

Besides the ordinary type of active, free-swimming fish, like the herring or trout or mackerel, there are fish flattened in adaptation to life on the bottom, some flattened sideways, like the sole and plaice, others flattened from above downwards, like the skates and rays; there are elongated fish like eels and pipe-fish; there are fish with prehensile tails, like the sea-horse; there are the flying-fish adapted for leaping long distances out of the water; mottled fish of irregular outline adapted for living on rocks; deep-sea fish with wonderful phosphorescent organs for searchlights and huge eyes for perceiving the faintest trace of light; cave-fish without eyes at all;

Sucker-fish adapted for sticking tight to the underside of stones, or for being carried about by larger fish without expending any energy themselves; and so on and so forth through almost every conceivable sort of form possible in an under-water existence. The examples previously mentioned give an idea of some of the adaptive radiation which has taken place in reptiles and in mammals; very similar instances could be taken from any other large group, such, for instance, as the insects.

It is interesting to take any such group and to see what part adaptive radiation has played in giving rise to the main sub-groups into which it is divided. There is one particular form of specialization which we have not so far mentioned; and that is the retrograde form of specialization known as degeneration.

There are many cases known where animals can be definitely shown to be less highly organized to-day than were their ancestors in the past. The animal known as Sacculina, for instance, is a parasite upon various sorts of crabs.

It consists of a mere bag filled with little else but reproductive cells, and sending out a whole series of branched roots which penetrate the crab's body in all directions and suck nourishment out of it. At first sight, the relationships of this unpleasant creature are very difficult to determine. But when its development is investigated, it is found that it hatches out of the egg as a a free-swimming larva exactly like those found in many Crustacea. It has jointed limbs, an external skeleton made of chitin, and in fact bears all the distinguishing marks that other Crustacea do, be they crabs or lobsters or shrimps or water-fleas.

It is, in fact, a crustacean which has become adapted to a parasitic life; and in so doing it has acquired special adaptations, such as the root-like organs, specially suited to that life, while it has lost sense-organs, limbs, digestive system, and everything else necessary for leading a free and independent existence.

It has lost more than it has gained, its organization has become simpler, its independence less; in fact it has gone down the evolutionary hill, and the direction of its history has been in most ways the opposite of the direction which characterizes biological progress. The form in which Sacculina hatches out resembles in general many crustacean larvae; but it is particularly like the early stages of the animals known as barnacles.

Every one who has been to the seaside knows what an acorn barnacle looks like—a little creature attached to rocks or piles, enclosed in a white shell, and capable of sending out of a slit in the top of the shell a regular sweep-net composed of a number of 'arms'—really appendages—with which it drags minute floating particles of food in towards its mouth. Even in the adult state a barnacle shows some resemblance to ordinary Crustacea, especially in its jointed limbs and chitinous external skeleton; the early free-swimming larva clinches the matter and gives complete proof, as in the case of Sacculina, of their crustacean nature and affinities.

The very close resemblance of the larva of barnacles to that of Sacculina points to an especially close connexion between the two sorts of animals; and as a matter of fact, the two are undoubtedly descended from a common stock. The barnacle is degenerate as well as the Sacculina, but it is not so degenerate. It still, for instance, possesses organs for capturing and digesting food. On the other hand, it has lost its organs of special sense and of locomotion. Further, it is not adapted to the same general mode of life as is Sacculina; it is not parasitic, but sedentary or sessile.

This settling down and becoming fixed is the other great cause, besides parasitism, of degeneration in animals; as would be expected, the degeneration due to a sedentary life is rarely so great as that due to parasitism, since the sedentary animal does not obtain its food ready-digested as do most parasites. It must not be supposed that because the general rule among animals is that time brings change, that therefore time *invariably* brings change. The common lamp-shell Lingula, for instance, has persisted without any appreciable change whatever in the structure of its shell for the prodigious period of time, certainly over five hundred million years, which has elapsed since the Cambrian epoch in the Primary period.

Even when individual species have changed, the general characters of groups have often persisted with very little modification, as, for instance, those of dragonflies since the coal-measures, of shark-like fish since the Silurian. In some cases this may mean that for some unknown reason the species or the group has lost the power of varying to any considerable extent.

More often, probably, it so admirably fills one particular niche in the order of things, and a niche which stays the same throughout the geological periods, that it pays for the animal or the group of animals to stay as they are, leaving it to other groups or to other branches of the same group to colonize new niches and to progress towards fuller existence. To sum up, we may say that two main types of evolutionary change result in animals (and also, as a matter of fact, in plants) from the struggle for existence and the constant appearance of inheritable variations.

In the first place, once a new type or plan of structure has been evolved, it undergoes adaptive radiation; in other words, there are developed a number of separate species all built on the same general ground-plan, but adapted to different and usually incompatible modes of life. In the second place, new types and new plans are continually appearing as time goes on, and progress is marked by the fact that among the later-evolved types there is to be found greater complexity of organization, greater control and independence of environment, than among the earlier. It might perhaps be thought that specialization was often the same thing as progress. Specialization, however, implies close adaptation to one particular mode of life, while progress means greater general efficiency.

If we look into the actual history of animals in the past, we find that specialization in one direction involves the sacrifice of possible advance in other directions, and is a barrier in the long run to any but a quite limited degree of progress.

As a result of its long course of specialization, extending for tens of millions of years through the better part of the Tertiary epoch, a specialization all tending towards greater efficiency in running and browsing, the horse stock has, it seems, cut itself off from the possibility of adapting itself to other modes of life—to a life in the water or in the trees, or to a carnivorous habit. There is a limit to the perfection which any line of specialization can attain.

While the horse was growing larger, developing hoofs, reducing the number of its digits, another and wholly different type was being evolved in the person of man. If it were not that horses are useful to man, and are accordingly domesticated, they would now be wholly or almost wholly extinct. The 'natural' enemies of the horse are the large carnivores. These are built on the same general plan as the horse—the mammalian—and indeed are the results of the specialization of the same plan in another direction. The same limits are thus set to them as to the horse stock.

Before the advent of man, a state of equilibrium existed between the horse and its enemies, the latter not able to destroy the former entirely, the first not able to escape the payment of some toll to the second. But the horse came up against the wholly new biological conditions introduced by the new type, man; it was in direct conflict with man's cunning and tools and his habit of hunting in bands; still more important, it had to compete with the indirect effects of the new development, such as the settlement and cultivation and enclosure of the land.

Against all this the horse was powerless. It could not develop far enough or fast enough to adapt itself to such sudden changes, and as a result it is becoming extinct as a wild species. Over and over again the same thing happens, and the specialized representative of the old type, plant or animal, is extinguished in competition with the specialized representative of the new. For tree-like representatives of the horsetail type, which existed in the Carboniferous period, we have seed-bearing trees to-day; for pterodactyls, birds and bats; for dinosaurs, the large mammals; for the large early amphibians, the Stegocephalia, we have crocodiles; for the abundance, both in numbers and in species, of the larger mammals in the Pliocene, we have the multifarious activities of the swarms of man.

The new type seems always to have arisen from some comparatively unspecialized branch of the old, and to have attained its pre-eminence by means of adaptations towards general instead of towards special efficiency.

Man has ousted the other mammals from their dominant position owing to the development of his mind. Through his particular type of mind he is able to deal rationally with the problems that confront him; tools, machines, tradition, civilization, and unexampled control and independence have been the result.

The human mind is not merely adapted for solving one or two particular problems; it represents a *method* more efficient than any previously adopted for dealing with any and every problem that may confront an organism. Man's body is not highly specialized, and he seems to have arisen, through a monkey-like ancestor, from some unspecialized early mammalian stock like the Insectivores.

Nor did the early mammals show any signs of specialization. All the fossil mammals that we know of during the time when the great period of reptilian dominance and specialization lasted were small, primitive creatures, at first sight not likely to wrest the palm from their powerful rivals. Very similar chains of events may be seen taking place in the evolution of human machines and inventions. Take, for example, the history of transport.

The most primitive method was the carrying of single loads by human beings or pack-animals. After that came the invention of vehicles in general and of wheeled vehicles in particular. The wheeled vehicle became specialized ('adaptive radiation') in innumerable ways.

We have the war-chariot; the rapid vehicle for passenger transport and for pleasure; the heavy wagon and cart; the van and pantechnicon; the four-inhand mail-coaches bowling daily at fixed hours along the main roads of the land.

A limit was set to the capacity and the speed of such vehicles by the speed and strength of beasts of burden on the one hand, and by the imperfections of road-surface on the other. At the beginning of last century a new type of vehicle was evolved for which these limitations no longer existed. It was discovered how to replace the energy of animals by that of steam, and in large part to overcome the difficulties of surface friction by making the wheels of the vehicles run on metal rails.

As a result, steam locomotives became for certain purposes the 'dominant type' of vehicle within an extremely short period of time. Here again the type itself has been improved, so that we have now for some time been close to the limit of its possibilities.

It does not seem possible to run a profitable service at speeds of much over sixty miles an hour. About half a century later a new plan was evolved. The internal combustion engine was produced, and gave certain great advantages, notably in being ready to start at once without the long preparation of 'getting steam up'.

It appears to be the fate of new types to lead a precarious existence for a considerable time before they can compete successfully with the dominant types of the period. This was so, as we saw, with the earliest mammals; it was so for the steam-engine; and it was so, to a very marked degree, for the internal combustion engines. They were laughed at when their inventors took them out on the roads; the law laid down that they should be preceded by a man with a red flag; the early defects in construction did actually occasion many a breakdown.

But within thirty years they came into their own. Meanwhile, still another competitor is in the field—the flying machine, a totally new type, abandoning not only the particular device of the wheel, but the whole element to which the wheel was adapted. It looks as if in certain respects, where speed is the main object, the aeroplane would become the dominant type of vehicle; but that it would leave the major part of transport to be dealt with by train and by motor.

Another interesting parallel with the evolution of organisms is found in the fact that although there has been progress, although the dominant type of vehicle has altered with the passage of time, yet many representatives of the old types have survived. They have managed to survive by becoming restricted to a few special conditions and places. Packanimals, for instance, while once universal, are now only employed in mountainous or roadless countries.

The American buggy is still of the greatest use over the unmade roads which are still to be found in so many parts of the United States. The horse-drawn vehicle will long hold its own in businesses in which not much capital is available, and speed and great power are not of the first importance. This survival of all or almost all the types that arise in evolution, even though new types arise and supremacy changes hands, is of general occurrence in the development of animals and plants, and is at first sight very puzzling. However, the insight which is gained by looking, as we have done, into the evolution of something familiar and human like the means of locomotion, helps us to understand the more complex and slower-moving processes of organic evolution.

Another point which is brought out by the study of the development of some human contrivance such as the means of transport, is the great speed of change now possible in human affairs as against the slowness of change prevailing in lower organisms. The whole period from the stage-coach to the aeroplane is comprised in well under two centuries. The resulting change in human habits has been enormous; to produce comparable changes in the habits of an animal stock there would be needed a period certainly to be reckoned in tens of thousands, possibly in millions of years.

On the other hand, the evolution of machines is a perfectly real evolution. Two different types of machines capable of performing the same general function—such, for instance, as the motor-lorry and the goods steam-engine—do come into a very real competition with each other, and the issue of the struggle is decided by a form of true natural selection, depending in the long run upon which of the two pays the better.

Here again the study of machines throws light upon the course of events in animals. It is often supposed that evolution must involve some conscious effort on the part of the evolving organism, that the struggle for existence is a conscious struggle, or that a species in some mysterious way 'learns' how to develop some new improvement in its structure. As a matter of fact, almost the whole of such ideas are purely metaphorical, and arise simply because we read the processes of our own minds into the operations of nature; it is not scientifically correct to speak, for instance, of *purpose* except in relation to human minds.

We see at once that the machines have no idea themselves of the direction of their evolution, that the 'struggle' between them is only a metaphorical struggle, that the selection between them is so far as they are concerned a mere sifting process, the issue of which depends upon the advantages or disadvantages which they may happen to possess. It is the same with organisms.

If two races of animals come into competition, the issue is decided by the qualities which each happens to possess; 'natural selection' is a name for the effect exerted by all the forces of the environment with which they come into relation, an effect which acts again like an automatic sieve, and lets some through to perpetuate themselves, keeps back and so extinguishes others.

The 'struggle' and the 'competition' are again usually metaphorical only. For instance, when the common housesparrow was introduced into America, it entered into competition with many of the small sparrow-like birds which had developed in that country. But the struggle did not take the form of a war between the invader and the original inhabitants. It was an indirect struggle, due to the fact that both lived upon the same sort of food, both occupied the same sort of sites. The European sparrow happened to be endowed with qualities which gave it an advantage, and as a result it has spread enormously over the American continent, while many of the native birds have correspondingly decreased.

If we wish to use a human metaphor, we can say that its success has been the result of 'peaceful penetration', not of fighting. The difference between machines and organisms, of course, is that the machines are directly designed by man, whereas the place of the designer in animal evolution is (roughly) taken by the variation which seems to be universal in organisms. It is variation which provides the raw differences upon which the sieve of natural selection can work. Mention must also be made of the theories of evolution which are summed up under the term *orthogenesis*, which means evolution in straight lines.

It is frequently found, as for instance in the development of the horses or of the elephants, that evolution as revealed by fossils proceeds straight onward through geological time in a perfectly definite direction-in the horses towards single toes and hoofs, in the elephants towards great bulk, tusks, and trunk. Orthogenesis is sometimes used merely as a descriptive term, to denote this observed fact of straight-line evolution. But by others it is used to mean that there exists some inner necessity for the evolutionary line in question to develop in just that one way and no other. However, a series of fossils, even if beautifully complete, can really give us no insight into the method by which its evolution occurred.

Whenever the direction in which the series is evolving is adaptive or seems biologically advantageous, the orthogenetic series can be perfectly well explained by natural selection. On the other hand, there do exist cases where at least no advantage can be perceived by us in the direction pursued. This is so, for instance, as regards the Ammonites, the extinct cephalopod molluscs which died out near the end of the Secondary period. Near the close of their time on earth, they often evolved orthogenetically into the most bizarre forms, their spiral shell becoming unwound, or irregular. Possibly in such cases a real causal orthogenesis, an inwardly determined mode of variation, is at work. But we are justified in saying that such cases, if they occur at all, are certainly rare.

Chapter 11

Animal Reproduction

Reproduction is one of the most characteristic features of living organisms. Life would not exist on Earth if plants and animals did not reproduce to make their offspring. By reproducing, a living organism can be sure that there is another individual of its kind to take its place when it dies. In this way a species of organism guarantees its survival.

A species is a particular type of organism. For example, a horse is a species and a zebra is another species. A species which cannot reproduce enough offspring will disappear for ever from the face of the Earth - it will become extinct. This has happened many times in the past. The best known example of animals which have become extinct is the dinosaurs.

The dinosaurs were a group of reptiles which mysteriously became extinct 60 million years ago.

Fortunately there has always been another type of living organism to replace those that become extinct. In the case of the dinosaurs they left the Earth to the group of animals that we belong to, the mammals.

How Animals Reproduce

Animals can be grouped into those which give birth to living offspring and those which lay eggs that eventually hatch into offspring. Those animals which give birth to live offspring are called live-bearing or viviparous.

Those animals which lay eggs are called egg-laying or oviparous. The difference is in the place where the offspring develops before it is born. Below is a table of the vertebrates which shows which groups are viviparous and which groups are oviparous.

Group	*Oviparous (Egg-layers)*	*Viviparous* (Live-bearers)
Mammals	Only a few primitive egg-laying mammals exist. They live in Australia and New Guinea. e.g. spiny anteater and duck-billed platypus.	Nearly all mammals e.g. mouse, human, cat, dog, bear, kangaroo and dolphin.
Birds	All birds. e.g. robin, penguin, parrot, sparrow and eagle.	None.
Reptiles	Most reptiles are egg-layers. e.g. crocodile, turtle and cobra.	Some lizards and snakes are live-bearing. e.g. vipers.
amphibians	Nearly all. e.g. frog, toad and	A few species of frogs living in South America and West Africa are live-bearers.
Fish	Most species. e.g. herring, salmon and trout.	Quite a few species are live-bearers. e.g. sharks and guppies.

Amongst the invertebrates there are many which are oviparous but a few are viviparous such as sea anemones and aphids.

Sexes and Sex Cells

A species of animal usually exists in two types or sexes called males and females. Each sex has its role to play in reproduction.

When an animal gives birth or lays eggs we notice that it is always the female which does this. In some cases, such as the sea horse and the midwife toad, the male appears to give birth to the offspring.

In these exceptions the male has been given the eggs to look after but it is still the female animal that lays the eggs in the first place. To produce an offspring two special cells are needed. A cell is a microscopic part of an organism's body. The bodies of animals are made of millions of cells. There are many different sorts of cells in a body, each having specialized functions.

Some are found in the blood transporting oxygen, others are found in the walls of our stomach producing juices to digest our food. To reproduce, animals make special sex cells. In the male animal these sex cells are called sperm cells or sperms and in the female animal they are called egg cells or eggs.

Sperm Cells

Sperm cells are very small but they are very specialized too.

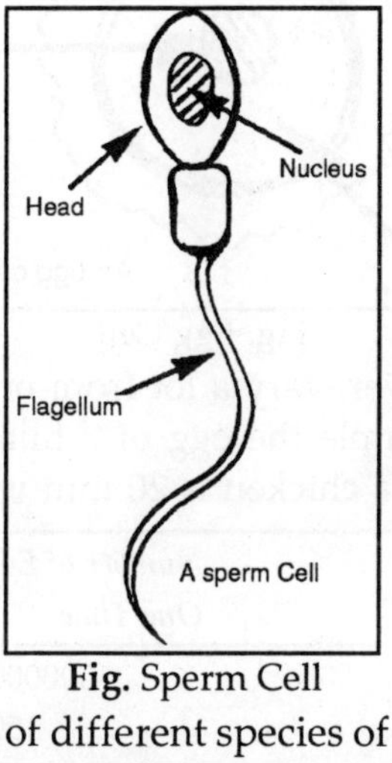

Fig. Sperm Cell

The sperm cells of different species of animal are all about the same size, about 60μm long (μm = micrometres; This is a thousandth of a millimetre). Over 11 000 sperm cells could fit on a pin head! All sperms have a head and a tail called a flagellum.

They use the flagellum to swim through liquids. The head of the sperm is very important because it contains the nucleus. The nucleus of the cell is the control centre. In sex cells the nucleus carries half of the information needed for reproduction (like the plans needed to construct a house).

Sperm cells are made in very large numbers by special organs in the male's body called testes (sing. testis). In most animals the testes are carried inside the male's abdomen but in mammals they are kept in a sack of skin called sperm duct the scrotum outside the abdomen.

Egg Cells

Egg cells are produced by female animals in special organs called ovaries (sing. ovary). These are found inside the abdomen of all female animals. Egg cells are much bigger and simpler than sperm cells in their structure.

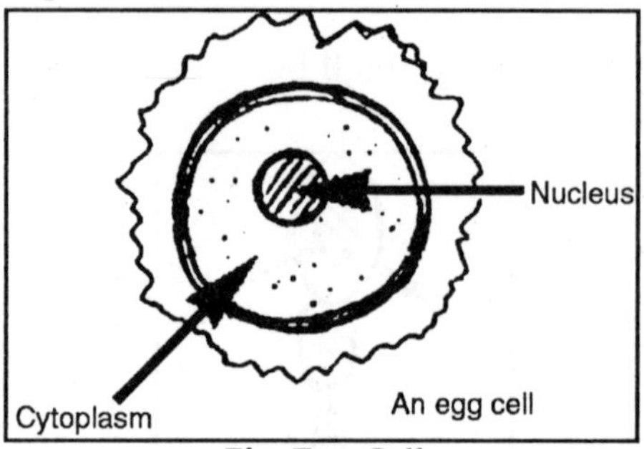

Fig. Egg Cell

They do, however, vary a lot from one species of animal to another. For example the egg of a human is only 0,1 mm wide but the egg of a chicken is 20 mm wide.

Animal	*Number of Eggs Produced At One Time*
Cod	5000000
Frog	1500
Viper	12
Human	1

Even so the human egg cell is still nearly 50 times wider than a sperm cell. The reason for the difference in size is that the egg has a large food supply stored inside it called yolk. Yolk is the yellow part of a chicken's egg.

The egg also has its own nucleus which carries the other half of the information needed for reproduction. Because eggs are bigger than sperms they are not produced in such large numbers. Even so animals which do not take care of their offspring, such as fishes and frogs, will lay a large number of eggs. Animals which take care of their offspring, such as mammals and birds, will produce less eggs.

MATING AND FERTILIZATION

The sperm cell carries half of the information needed for reproduction and the egg cell carries the other half. To produce an offspring the sperm and egg have to meet. This means that the male and female animals have to come together. This is called mating. Because the sperms can swim on their own, aquatic animals such as fish can simply release the eggs and sperms into the water. Amphibians, which return to water to reproduce, also release their eggs and sperms directly into the water.

The sperm cell is microscopic and only lives for a short time, so the eggs and sperms need to be released close together. Fish, such as herring, which live in the open sea will shoal at certain times of the year. Shoaling occurs when the males and females gather close together, usually near the surface of the sea. The females release their eggs and the males release their sperms at the same time and in the same place. Therefore, the sperms do not have far to swim.

Fish, such as salmon, trout and sticklebacks, build their nests in the mud and gravel on the bottom of shallow rivers and lakes. The male usually builds the nest and then he guards it. When a female enters his territory he will try to encourage her to lay her eggs in his nest. This sometimes involves quite a complicated mating display. If the female decides to lay her eggs in his nest he will release his sperms over the eggs.

Amphibians

Animals which live on land have a problem sperm cells cannot walk or fly! For the amphibians the answer to this problem is to return to the water to mate. Frogs and toads will migrate to the ponds and lakes where they were born. They do this usually in Spring. Sometimes they cross roads on their migration route in such large numbers that they can cause accidents! In France the motorway which runs through the Sologne, south of Orleans, has special tunnels constructed underneath it.

These tunnels allow the amphibians, which are very numerous in this region, to cross underneath the road safely. These tunnels are called crapauducts! When the frogs arrive at the water the males and females call to one another. This helps the amphibians to identify one another, as many different species may be mating at the same time and in the same place.

The male holds the female round the waist so that as she releases her eggs he will release his sperms close to them in the water. The mating calls are very important for some species because the males will try to mate with any object floating in the water, including fish or pieces of wood!

Mating on Land

Vertebrate animals such as reptiles, birds and mammals, and invertebrates, such as spiders and insects, do not return to water to mate, so the male must place the sperm cells directly inside the female's body. The male often has a special organ to place the sperms inside the female and the female has a special opening to receive the sperms. This kind of mating is called copulation and it has two advantages. First, the sperms are released in a liquid called semen which is made by the male. The sperms can swim in this liquid towards the eggs inside the female.

The second advantage is that the sperms cells inside the female are more certain to meet the egg cells. Animals which mate in this way do not produce so many eggs because only a few of them are lost. Copulation is so effective that many aquatic animals use it too, even though they are surrounded by water in which the sperms could swim. Getting closer to the female can have its disadvantages. If the animal is a carnivore, such as a spider, the female may think that the male is a meal to be eaten rather than a mate!

The result is that land animals have developed complicated mating behaviour which involves dances, smells and often very colourful mating displays. It is usually the male which is colourful in the animal world, even amongst the herbivores. This is very apparent amongst birds; the peacock, for example, has brilliant tail feathers. The female (peahen) by comparison is quite drab.

Fertilization

When a sperm cell meets an egg cell, the head of the sperm fuses with the egg. The flagellum of the sperm is lost, its work is finished. The information in the nucleus of the male's sperm can combine with the information in the nucleus of the female's egg to produce an offspring. This important event is called fertilization. If fertilization takes place inside the female's body - as it does in mammals or birds - it is called internal fertilization. Only one sperm cell will fertilize one egg cell successfully. It is possible, however, for two sperms to fertilize two eggs.

Species	*Litter Size*
Bat	1
Dolphin	1
Chimpanzee	1
Lion	3
Hedgehog	5
Red Fox	6
Rabbit	6
Black rat	11

This will result in two offspring. This is not at all unusual in the animal world. Foxes give birth to litters of between four and eight puppies. A litter is a group of babies born from the same mother at the same time. The table below gives some examples of different mammals and their average litter size

Reproduction in Birds

The egg cell of a bird is really what we call the yolk. In the chicken's egg this yellow ball of yolk is about 20 mm in diameter. The yolk is the food supply for the developing embryo. The nucleus of the bird's egg can be found in the middle of a white spot which is found on top of the yolk. You need a microscope to see the nucleus itself but the white spot is visible to the naked eye. The egg that a hen (a female bird) lays is covered in layers of albumen (a material like jelly), two membranes and a hard shell. The shell and the membranes protect and support the egg.

The albumen feeds the developing embryo with water and protein. The albumen, membranes and shell are added to the yolk layer-by-layer as it passes down a tube called the oviduct inside the hen's body. The oviduct carries the egg from the ovary out of the hen's body. The last layer to be added is the shell. If a bird's egg is to be fertilized, mating must take place so that the egg will meet a sperm cell in the oviduct before these layers are added. The shell of a bird's egg is often coloured or patterned to help camouflage it from predators.

This is especially important for birds which nest on the ground, such as plovers. A hen bird may lay from 1 to 19 eggs in her nest. A group of eggs laid by a hen is called a clutch. It does not matter whether the hen bird lays 2 or 20 eggs in her clutch, she will only lay one egg per day. This is important because birds' eggs are very large. If the hen bird were to lay all her eggs at once she would have to produce smaller eggs or fewer eggs. The hen bird usually lays her egg for the day early in the morning. This is so that she can be free to hunt for food during the morning and in the middle of the day.

Taking Care of the Eggs

Birds take great care of their eggs. Usually the eggs are laid in a nest. Sometimes the nest can be a simple scrape in the ground or it can be a very elaborate nest. For example the mallee fowl of eastern Australia builds a huge nest of decomposing vegetation which helps to keep the eggs warm.In most cases it is the body heat of the parent bird which keeps the eggs warm. Keeping the eggs warm is called incubation or brooding. If the eggs get cold the embryo developing inside will die.

In some types of birds the hen develops a patch on her breast without feathers. The eggs that she is incubating will be in contact with the skin and so they will be close to her body temperature. The hen must go and feed from time to time, so the cock (the male bird) may take over. This is particularly important for one species of bird, the emperor penguin. This is the largest species of penguin, it has a mass of 16 kilogrammes and it lives on the continent of Antarctica. Antarctica is mostly covered in solid ice but it is there that the emperor penguin breeds.

It is the male bird that looks after the egg to begin with. The hen lays only one large egg per year. She uses up a lot of her food reserves to make this egg, so she must go to sea to fish for food to replace these reserves. The male stays on the ice and keeps the egg on his feet covered by a fold of skin. Obviously the emperor penguins cannot build a nest from the ice, so the male keeps the egg on his feet for eight weeks through the fierce Antarctic winter.

The temperatures can drop to - 60° C. The chick hatches at about the time when the hen returns with food. At last the cock can go and hunt for food. He will return about two weeks later with more food for the chick.

Reproduction of Horse

Pregnancy lasts for approximately 335-340 days and usually results in one foal (male: colt, female: filly). Twins are rare. Colts are usually carried 2-7 days longer than fillies. Females 4 years and over are called mares and males are stallions. A castrated male is a gelding. Horses, particularly colts, may sometimes be physically capable of reproduction at approximately 18 months but in practice are rarely allowed to breed until a minimum age of 3 years, especially females. Horses four years old are considered mature, though the skeleton usually finishes developing at the age of six, and the precise time of completion of development also depends on the horse's size (therefore a connection to breed exists), gender, and the quality of care provided by its owner.

Also, if the horse is larger, its bones are larger; therefore, not only do the bones take longer to actually form bone tissue (bones are made of cartilage in earlier stages of bone formation), but the epiphyseal plates (plates that fuse a bone into one piece by connecting the bone shaft to the bone ends) are also larger and take longer to convert from cartilage to bone as well.

These plates convert after the other parts of the bones do but are crucial to development. Depending on maturity, breed and the tasks expected, young horses are usually put under saddle and trained to be ridden between the ages of two and four.

Although Thoroughbred and American Quarter Horse race horses are put on the track at as young as two years old in some countries, horses specifically bred for sports such as show jumping and dressage are generally not entered into top-level competition until a minimum age of four years old, because their bones and muscles are not solidly developed, nor is their advanced training complete. For endurance riding competition, horses may not compete until they are a full 60 calendar months old. Horses have, on average, a skeleton of 205 bones.

A significant difference in the bones contained in the horse skeleton, as compared to that of a human, is the lack of a collarbone—their front limb system is attached to the spinal column by a powerful set of muscles, tendons and ligaments that attach the shoulder blade to the torso. The horse's legs and hooves are also unique, interesting structures.

Their leg bones are proportioned differently from those of a human. For example, the body part that is called a horse's "knee" is actually the carpal bones that correspond to the human wrist. Similarly, the hock, contains the bones equivalent to those in the human ankle and heel.

The lower leg bones of a horse correspond to the bones of the human hand or foot, and the fetlock (incorrectly called the "ankle") is actually the proximal sesamoid bones between the cannon bones (a single equivalent to the human metacarpal or metatarsal bones) and the proximal phalanges, located where one finds the "knuckles" of a human.

A horse also has no muscles in its legs below the knees and hocks, only skin and hair, bone, tendons, ligaments, cartilage, and the assorted specialized tissues that make up the hoof.

Reproduction of Dog

In domestic dogs, sexual maturity (puberty) begins to happen around age 6 to 12 months for both males and females, although this can be delayed until up to two years old for some large breeds. Adolescence for most domestic dogs is around 12 to 15 months, beyond which they are for the most part more adult than puppy. As with other domesticated species, domestication has selectively bred for higher libido and earlier and more frequent breeding cycles in dogs, than in their wild ancestors.

Dogs remain reproductively active until old age. Most female dogs have their first estrous cycle between 6 and 12 months, although some larger breeds delay until as late as 2 years. Females experience estrous cycles biannually, during which her body prepares for pregnancy, and at the peak she will come into estrus, during which time she will be mentally and physically receptive to copulation.

Dogs bear their litters roughly 56 to 72 days after fertilization, although the length of gestation can vary. An average litter consists of about six puppies, though this number may vary widely based on the breed of dog. Toy dogs generally produce from one to four puppies in each litter, while much larger breeds may average as many as 12 pups in each litter.

Spaying and Neutering

Neutering (spaying females and castrating males) refers to the sterilization of animals, usually by removal of the male's testicles or the female's ovaries and uterus, in order to eliminate the ability to procreate, and reduce sex drive. Neutering has also been known to reduce aggression in male dogs, but has been shown to occasionally increase aggression in female dogs.

Animal control agencies in the United States and the ASPCA advise that dogs not intended for further breeding should be neutered so that they do not have undesired puppies. Because of the overpopulation of dogs in some countries, puppies born to strays or as the result of accidental breedings often end up being killed in animal shelters.

Neutering can also decrease or eliminate the risk of hormone-driven diseases such as mammary cancer, as well as undesired hormone-driven behaviours. However, certain medical problems are more likely after neutering, such as urinary incontinence in females and prostate cancer in males.

The hormonal changes involved with sterilization are likely to somewhat change the animal's personality, however, and some object to neutering as the sterilization could be carried out without the excision of organs.

It is not essential for a female dog to either experience a heat cycle or have puppies before spaying, and likewise, a male dog does not need the experience of mating before castration.

Female cats and dogs are seven times more likely to develop mammary tumors if they are not spayed before their first heat cycle. The high dietary estrogen content of the average commercial pet food as well as the estrogenic activity of topical pesticides may be contributing factors in the development of mammary cancer, especially when these exogenous sources are added to those normal estrogens produced by the body.

Dog food containing soybeans or soybean fractions have been found to contain phytoestrogens in levels that could have biological effects when ingested longterm.

Gender-preservative surgeries such as vasectomy and tubal ligation are possible, but do not appear to be popular due to the continuation of gender-specific behaviours and disease risks.

According to the Humane Society of the United States, 3-4 million dogs and cats are euthanized each year in the United States and many more are confined to cages in shelters because there are many more animals than there are homes.

Spaying or castrating dogs helps keep overpopulation down. Local humane societies, SPCAs and other animal protection organizations urge people to neuter their pets and to adopt animals from shelters instead of purchasing them.

Several notable public figures have spoken out against animal over population, including Bob Barker.

On his game show, The Price Is Right, Barker stressed the issue at the end of every episode, saying: "Help control the pet population. Have your pets spayed or neutered." Dogs are susceptible to various diseases, ailments, and poisons, some of which affect humans in the same way, others of which are unique to dogs.

Dogs, like all mammals, are also susceptible to heat exhaustion when dealing with high levels of humidity and/or extreme temperatures. Infectious diseases commonly associated with dogs include rabies (hydrophobia), canine parvovirus, and canine distemper.

Inherited diseases of dogs can include a wide range from elbow or hip dysplasia and medial patellar luxation to epilepsy and pulmonic stenosis.

Canines can get just about anything a human can get (excluding many infections which are species specific) like hypothyroidism, cancer, dental disease, heart disease, etc. Two serious medical conditions affecting dogs are pyometra, affecting unspayed females of all types and ages, and bloat, which affects the larger breeds or deep chested dogs.

Both of these are acute conditions, and can kill rapidly; owners of dogs which may be at risk should learn about such conditions as part of good animal care. First generation hybrids (such as this terrier mix) often are healthier than either parent due to the genetic phenomenon of heterosis or "hybrid vigor". Common external parasites are various species of fleas, ticks, and mites.

Internal parasites include hookworms, tapeworms, roundworms, and heartworms. See also CVBD (Canine Vector-Borne Diseases). Some breeds of dogs are also prone to certain genetic ailments, such as hip dysplasia, luxating patellas, cleft palate, blindness, or deafness.

Dogs are also susceptible to the same ailments that humans are, including diabetes, epilepsy, cancer, and arthritis. Gastric torsion and bloat is a dangerous problem in some large-chested breeds.

The typical lifespan of dogs varies widely among breeds. Based on questionnaire surveys of owners in the UK, Denmark, and the USA/Canada, the median longevity of most dog breeds is between 10 and 13 years.

The breed with the dubious distinction of the shortest lifespan (among breeds for which there is a questionnaire survey with a reasonable sample size) is the Dogue de Bordeaux with a median longevity of about 5.2 years, but several breeds, including Miniature Bull Terrier, Bulldog, Nova Scotia Duck Tolling Retriever, Bloodhound, Irish Wolfhound, Greater Swiss Mountain Dog, Great Dane, and Mastiff, are nearly as short-lived, with median longevities between 6 and 7 years.

On the other end of the spectrum, the longest-lived breeds, including Toy Poodle, Border Terrier, Miniature Dachshund, Miniature Poodle, and Tibetan Spaniel, have median longevities between 14 and 15 years. The median longevity of mixed breed dogs (average of all sizes) is one or more years longer than that of purebred dogs (all breeds averaged). As a rule of thumb, small breeds are longer-lived than large breeds, but some of the longest lived large breeds have median longevities nearly as large as those of the shortest lived small breeds, and some of the breeds with the shortest longevities are medium-sized.

"Median longevity" refers to the age at which half the dogs in a population have died and half are still alive. Individual dogs, even in breeds with low median longevities, may live well beyond the median. The dog widely reported to be the longest-lived on record is "Bluey," purportedly born in 1910 in Australia. He died in 1939 at the age of 29.5 years.

Bluey is usually identified as an Australian Cattle Dog, but the first Australian Cattle Dog breed standard was written in 1902, only eight years before Bluey's birth. It is unclear how closely Bluey was related to the breed as it exists today. The Bluey record is anecdotal and unverified. The longest verified records are of dogs living to 24 years.

Chapter 12

Breeding of Animals

The science of Animal breeding is defined as the application of the principles of Genetics and biometry to improve the efficiency of production in farm animals. These principles were applied to change animal populations thousands of years before the sciences of genetics and biometry were formally established.

The practice of animal breeding dates back to the Neolithic period (approximately 7000 BC), when people attempted to domesticate wild species such as reindeer, goats, hogs and dogs. Domestication was performed through controlled mating and reproduction of captive animals which were selected and mated based on their behaviour and temperament. Judging from cave paintings that have survived, selection was also applied to some qualitative traits such as coat colour and the absence or presence of horns.

Without written records, there is no certain knowledge of the evolution of animal breeding practices, but written documents dating back more than 4000 years indicate that humans appreciated the significance of family resemblance in mating systems, recognized the dangers of intense inbreeding, and used castration to prevent the reproduction of undesirable males.

Progress in the performance of domesticated animals through these selection practices was very slow; improvements were mainly due to animals adapting better to their environments. Robert Bakewell, an English animal breeder of the 18th century, is considered the founder of systematized animal breeding. He was the first to emphasize the importance of accurate breeding records, introduced the concept of progeny testing to evaluate the genetic potentials of young sires, and applied inbreeding to stabilize desired qualitative traits. He also promoted concepts such as "like begets like," "prepotency is associated with inbreeding" and "breed the best to the best."

Bakewell and his contemporaries in Europe pioneered the development of diverse breeds of Beef Cattle, Dairy Cattle, Sheep, Hogs and Horses. Most livestock breeds with pedigree herd books and breed associations were established between the late 18th century and the second half of the 19th century.

Colour, conformation, geographical origin and some production characteristics were the main factors that differentiated these breeds. Wide geographical redistribution of animal populations was also an important factor in the formation of new breeds, as invading armies, migrating people and traders transported livestock to new lands.

Animal breeding as a modern science belongs to the 20th century. Although numerous geneticists and biometricians have made significant contributions to the development of this science, J.L. Lush of Iowa State University is considered the father of the modern science of animal breeding. Lush and his students developed major scientific procedures applicable to the genetic improvement of farm animals.

SELECTIVE BREEDING

Selective breeding in domesticated animals is the process of a breeder developing a cultivated breed over time, and selecting qualities within individuals of the breed that will be best to pass on to the next generation. Breeding techniques such as inbreeding, linebreeding and outcrossing are utilized by breeders in the maintenance and improvement of their chosen breeds.

Breeding Stock

"Breeding stock" is a term used to describe a group of animals used for purpose of planned breeding. When an individual is looking to breed animals, he or she is looking for certain valuable traits in purebred stock for a certain purpose, or may intend to use some type of crossbreeding to produce a new type of stock with different, and presumably superior abilities in a given area of endeavor. For example, to breed chickens, a typical breeder intends to receive eggs, meat, and new, young birds for further reproduction.

Thus the breeder has to study different breeds and types of chickens and analyze what can be expected from a certain set of characteristics before he or she starts breeding them. Accordingly, when purchasing initial breeding stock, the breeder seeks a group of birds that will most closely fit the purpose intended.

Purebred Breeding

The idea of *breed purity* may strike an unpleasant chord with some individuals because it is reminiscent of nineteenth-century eugenics notions of a "superior strain" in humans, supposedly exemplified by members of an aristocratic social class or of certain races. The application of theories of eugenics had far-reaching consequences for human beings, particularly in cases such as The Holocaust. However, the idea of a superior strain in animals, such as that incorporated in the breeding of Thoroughbred race horses, was less controversial.

By "breeding the best to the best," employing a certain degree of careful inbreeding, considerable culling, and selection for "superior" qualities, one could develop a bloodline or "breed" superior in certain respects to the original base stock or landrace which had been produced primarily by natural selection.

This process forms the basis for a "true-breeding" line of animals, which today is generally called a "purebred" line. Once created, a purified line is "closed" and preserved from dilution and possible "debasement" by unrelated stock. When such animals are recorded with a breed registry, the closing of a bloodline group is sometimes referred to as having a closed stud book. However, this theory is not without some negative consequences.

When the ideal of purified lineage or purely aesthetic type is seen as an end in itself, without consideration of usefulness or vigor, a breed can suffer over time. Also, if a breed registry is closed with too small a gene pool, over time, degradation due to inbreeding depression may occur. For many purebred breeders, the need for a periodic outcross is recognized and incorporated into an overall plan.

Some breed registries thus maintain an "open" stud book for certain types of approved outcrossed or crossbred animals. The observable phenomenon of hybrid vigor stands in contrast to the notion of breed purity. However, on the other hand, indiscriminate breeding of crossbred or hybrid animals may also result in degradation of quality, particularly when animals are mass-produced without concern for overall health or longevity.

Backyard Breeding

The term backyard breeder is a general term, sometimes considered derisive, used to describe people who are "home breeders", and breed only a few animals for personal use or occasional sale. However, while some backyard breeders pay strict attention to quality, the term usually describes those who allow animals, particularly dogs or horses, to procreate regardless of physical, genetic, and/or emotional health.

It is also used to describe unreputable commercial breeders. In the process of careless breeding, many backyard breeders produce genetically weak animals that can be predisposed to genetic disorders or debilitating physical deformities, particularly those that result from inbreeding.

In dog breeding, these individuals focus on only a limited aspect of the dog, while ignoring the original function and temperament of the breed. When such breeding is carried out on a large scale, the venue is called a *puppy mill* or *puppy farm*. Backyard breeding is popularly blamed for the proliferation of aggressive dogs for the sports of dog baiting and dog fighting.

Dog fanciers generally believe that such ill-bred dogs are the reason for the bad public reputation of certain dog breeds, and the increasingly frequent enactment of breed-specific legislation. Because of the time and expense of feeding and caring for horses, which produce one foal per year, there are fewer horse breeders who produce animals *en masse,* though some individuals do engage in animal hoarding and breed far more animals than they can support.

In the horse world, overbreeding of grade animals that cannot be sold except for very low prices raises concerns that such animals will be slaughtered for horsemeat. Some, though not all, colour breeders make errors similar to puppy mill breeders by focusing on only one characteristic to the detriment of health, soundness, and quality conformation.

In addition, industrial breeders, such as those who produce *pregnant mare urine* for the manufacture of premarin, allow many mares to be bred indiscriminately, simply so that they are able to produce the urine laced with reproductive hormones that forms the base for the drug. Their foals are generally considered an unwanted byproduct, particularly males (colts), and sold off, frequently to the slaughter market.

IN-BREEDING AND OTHER BREEDING METHODS

One of the most hotly talked about topics with regard to pure-bred dogs is the use of in-breeding. This is a term that is often misused and is extremely misunderstood. Part of the misunderstandings come from differences in the way the terms are used within the scientific/medical field, and how it is commonly used by breeders. These are the most commonly accepted definitions used by serious dog breeders and will be the definitions used within this article.

In-breeding

This is the breeding of closely related animals. Brother-Sister, Parent-Offspring, ½ brother - ½ Sister.

Line-breeding

This is the breeding of animals that share common ancestors but are not closely related. For example the dogs may share a common great-grandparent.

Out-cross

This is generally considered the breeding of animals with no common ancestors within the first 4 or 5 generations.

COMMON MISCONCEPTIONS

In-breeding Causes Genetic Diseases

Breeding closely related animals increases the possibility that any bad genes in a line will show up. It does not 'cause' genetic disease.

Out-crossed Dogs are Healthier

This is only partly true. There is a known phenomenon called Hybrid Vigor. Two animals of unrelated strains breed and the offspring is often bigger and grows faster than it's purebred cousins.

This method is often used by farmers in order to get their animals to market sooner. But one of the biggest misconceptions of hybrid vigor is that it applies to all animals of mixed heritage. Hybrid Vigor only applies to the animals that are the direct offspring of the crossing of the unrelated strains.

In other words if you continue to breed animals of different strains there generally will not be any additional increase in hybrid vigor. If the unrelated strains share common genes for genetic disorders, hybrid vigor will not over ride the risk of the disorder showing up. Out-crossing can also cause problems if widely divergent physical types are mixed due to differences in growth rates and bone and muscle sizes.

BENEFITS OF EACH TYPE OF BREEDING

By definition, purebred dogs have a smaller gene pool to draw on than mixed breed dogs. That smaller gene pool gives the breed its individual characteristics, such as physical appearance and temperament. It is what makes a poodle a Poodle and a Golden Retriever a Golden Retriever. But there is considerable controversy with regard to whether the gene pools of the modern pure-bred have become too small.

Inbreeding

In-breeding is more likely to help "set" or "fix" a particular trait within a breed or a line by narrowing the gene pool to favour those traits. So if a breeder is looking to set a particular desirable feature of their line then in-breeding and choosing the offspring most strongly possessing that trait can be beneficial. In the first pedigree Sydney is the maternal son of Annie, thus doubling on Annie's genes.

Or another way to look at it is Annie is both Sydney's mother and Grand-mother. In the second example we have a full brother/sister in-breeding so we are doubling up on the genes from both Kiley and Annie. In-breeding can also help identify those bad genes that exist within a line. Dogs possessing the bad genes can be eliminated from a breeding programme and carriers also identified. Intermittent in-breeding within a line or breed is not damaging to the long term health of the animals.

However, in-breeding over successive generations can lead to reduced fitness and fertility problems among the offspring, resulting in a phenomena known as In-breeding Depression. It can take many generations to show up depending on the traits involved.

To use this method responsibly a breeder would not want to in-breed on animals with known genetic disorders, temperaments not in keeping with it's given breed, or known serious structural faults, or to in-breed frequently even on healthy-superior specimens.

Line-Breeding

Line-breeding is another way to help "set" or "fix" desirable traits. With line-breeding you breed animals that are related, but you are also routinely introducing genes from other lines into the genetic mix.

It takes longer to fix the desirable traits this way, but doing so lowers the risk of those problems associated with repeated in-breeding. With a tight line-breeding you might find the same 3, 4 or more dogs showing up numerous times in a 5 generation pedigree.

Here's an example of a tightly line-bred pedigree.

Loose line-breeding over successive generations will result in more variations of physical appearance than would in-breeding or tight line-breeding, but will keep the physical look and structure within the same general size and shape, it also carries fewer long term risks.

According to geneticists

Line-breeding can be carried on for many many generations without deleterious effects on the line or breed as long as the individuals involved have few hidden genetic disorders.

Out-Crossing

Out-crossing in terms of pure-bred dogs is the breeding of unrelated dogs. On a pedigree no names will be repeated within the first 5 generations. This type of breeding has both advantages and disadvantages.

Which as it turns out are flip sides of the same argument. With out-crossing you are maintaining the greatest genetic diversity, but this also leads to the least consistency in terms of physical appearance and other traits. Out-crossing does not guarantee that the animals won't develop genetic disorders, but it does tend to reduce the numbers of affected offspring.

Your best chance of getting an animal that is less prone to developing a genetic disorder comes more from finding a conscientious breeder that screens their animals for hereditary disorders and breeds for the betterment of the breed. All three methods of breeding have their place in a long term, well thought out breeding programme. Talk to the breeder, ask questions as to what their goal is in doing a particular breeding. Ask about the risks and what problems are known to that line. And all lines have some because the perfect dog and the perfect lines are still goals of the future for all breeders.

DOG

The dog is a domesticated subspecies of the wolf, a mammal of the Canidae family of the order Carnivora. The term encompasses both feral and pet varieties and is also sometimes used to describe wild canids of other subspecies or species.

The domestic dog has been one of the most widely-kept working and companion animals in human history, as well as being a food source in some cultures. There are estimated to be 400,000,000 dogs in the world. The dog has developed into hundreds of varied breeds.

Height measured to the withers ranges from a few inches in the Chihuahua to a few feet in the Irish Wolfhound; colour varies from white through grays to black, and browns from light to dark in a wide variation of patterns; and, coats can be very short to many centimeters long, from coarse hair to something akin to wool, straight or curly, or smooth. In breeding circles, a male canine is referred to as a *dog*, while a female canine is called a *bitch*.

The father of a litter is called the *sire*, and the mother of a litter is called the *dam*. Offspring are generally called *pups* or *puppies* until they are about a year old. A group of offspring is a *litter*. The process of birth is *whelping*. Many terms are used for dogs that are not purebred.

Origin and Evolution

Based on DNA evidence, the wolf ancestors of modern dogs diverged from other wolves about 100,000 years ago, and dogs were domesticated from those wolf ancestors about 15,000 years ago.

This date would make dogs the first species to be domesticated by humans. Evidence suggests that dogs were first domesticated in East Asia, possibly China, and some of the peoples who entered North America took dogs with them from Asia. As humans migrated around the planet a variety of dog forms migrated with them.

The agricultural revolution and subsequent urban revolution led to an increase in the dog population and a demand for specialization.

These circumstances would provide the opportunity for selective breeding to create specialized working dogs and pets.

This ancient mosaic, likely Roman, shows a large dog with a collar hunting a lion.

Molecular systematics indicate that the domestic dog descends from one or more populations of wild wolves. As reflected in the nomenclature, dogs are descended from the wolf and are able to interbreed with wolves. The relationship between human and canine has deep roots. Converging archaeological and genetic evidence indicate a time of domestication in the late Upper Paleolithic close to the Pleistocene/Holocene boundary, between 17,000 and 14,000 years ago.

Fossil bone morphologies and genetic analysis of current and ancient dog and wolf populations have not yet been able to conclusively determine whether all dogs descend from a single domestication event, or whether dogs were domesticated independently in more than one location. Domesticated dogs may have interbred with local populations of wild wolves on several occasions (a process known in genetics as introgression).

The earliest dog fossils, two crania from Russia and a mandible from Germany, date from 13,000 to 17,000 years ago. Their likely ancestor is the large northern Holarctic wolf, *Canis lupus lupus*. Remains of smaller dogs from Mesolithic cave deposits in the Middle East, dated to around 12,000 years ago, have been interpreted as descendants of a lighter Southwest Asian wolf, *Canis lupus arabs*. Rock art and skeletal remains indicate that by 14,000 years ago, dogs were present from North Africa across Eurasia to North America.

Dog burials at the Mesolithic cemetery of Svaerdborg in Denmark suggest that in ancient Europe dogs were valued companions. Genetic analyses have so far yielded divergent results. The ancestors of dogs split off from other wolves between 75,000 and 135,000 years ago, while a subsequent analysis indicated a "common origin from a single gene pool for all dog populations" between 40,000 and 15,000 years ago in East Asia.

Both sets of dates must be reevaluated in light of recent findings showing that poorly calibrated molecular clocks have systematically overestimated the age of geologically recent events. On balance, and in agreement with the archaeological evidence, 15,000 years ago is the most likely time for the wolf-dog divergence.

Development of Dog Breeds

Dogs have been bred into a variety of shapes, colours and sizes. Variation can be wide even within a breed, as with these Cavalier King Charles Spaniels. There are numerous dog breeds, with over 800 being recognized by various kennel clubs worldwide.

Many dogs, especially outside the United States of America and Western Europe, belong to no recognized breed. A few basic breed types have evolved gradually during the domesticated dog's relationship with humans over the last 10,000 or more years, but all modern breeds are of relatively recent derivation.

Many of these are the product of a deliberate process of artificial selection. Because of this, some breeds are highly specialized, and there is extraordinary morphological diversity across different breeds. Despite these differences, dogs are able to distinguish dogs from other kinds of animal. The definition of a dog breed is a matter of some controversy.

Depending on the size of the original founding population, closed gene pool breeds can have problems with inbreeding, specifically due to the founder effect. Dog breeders are increasingly aware of the importance of population genetics and of maintaining diverse gene pools.

Health testing and new DNA tests can help avoid problems, by providing a replacement for natural selection. Without selection, inbreeding and closed gene pools can increase the risk of severe health or behavioural problems. Some organizations define a breed more loosely, such that an individual may be considered of one breed as long as 75 per cent of its parentage is of that breed.

These considerations affect both pets and the show dogs entered in dog shows. Even prize-winning purebred dogs sometimes possess crippling genetic defects due to founder effect or inbreeding. These problems are not limited to purebred dogs and can affect cross-breed populations.

The behaviour and appearance of a dog of a particular breed can be predicted to a degree, while mixed-breed dogs show a broader range of innovative appearance and behaviour. Mixed-breed dogs or Mongrels (also called "mutts") are dogs that do not belong to specific breeds, being mixtures more than two in variant percentages.

Mixed breed dogs and purebred dogs are both suitable as companions, pets, working dogs, or competitors in dog sports. Sometimes different breed dogs are deliberately bred, to create cross-breeds such as the Cockapoo, a mixture of Cocker Spaniel and Miniature Poodle.

Such deliberate crosses may display some degree of hybrid vigor and other desirable traits, but may or may not inherit any of the desired traits of their parents, such as temperament or a particular colour or coat. Without genetic testing of the parents, the crosses can end up inheriting genetic defects that occur in both parental breeds.

A breed is a group of animals that possesses a set of inherited characteristics that distinguishes it from other animals within the same species. Deliberately crossing two or more breeds is also a manner of establishing new breeds, but it is only a breed when offspring will reliably demonstrate that particular set of characteristics and qualities.

Breed Popularity

Breed popularity varies widely over time and in different parts of the world and different segments of the population. Counting by American Kennel Club (AKC) registration (not by licensing registration or by United Kennel Club (UKC) registration, which could present different statistics), the Labrador Retriever has been the

United States's most commonly registered breed of dog since 1991. However, even within parts of the United States, popularity varies; for example, in 2005 the most-registered breed in New York City was the Poodle while the Yorkshire Terrier was the second-most-registered breed in Houston.

However, animal shelters in many parts of the United States report that the most-commonly available dog for adoption is the American Pit Bull Terrier or pit bull-type mixes, making up as much as 20 per cent of dogs available for adoption, none of which would be registered with the AKC. Two decades ago, in 1983, the AKC's top two registered breeds were the American Cocker Spaniel and the Poodle.

In the United Kingdom, The Kennel Club reports that the most-registered breed from at least 1999 to 2005 was the Labrador Retriever. It rounds out the top three for 1999 to 2005 with the German Shepherd Dog, also popular in the States, and the English Cocker Spaniel, which is no longer even in the top ten in the States. In the United Kingdom a national dog adoption and rescue service indicates that the most common breed appearing in shelters is the Greyhound followed by the Staffordshire Bull Terrier.

Physical Characteristics

Modern dog breeds show more variation in size, appearance, and behaviour than any other domestic animal. Within the range of extremes, dogs generally share attributes with their wild ancestors, the wolves. Dogs are predators and scavengers, possessing sharp teeth and strong jaws for attacking, holding, and tearing their food.

Although selective breeding has changed the appearance of many breeds, all dogs retain basic traits from their distant ancestors. Like many other predatory mammals, the dog has powerful muscles, fused wristbones, a cardiovascular system that supports both sprinting and endurance, and teeth for catching and tearing. Compared to the bone structure of the human foot, technically, dogs walk on their toes.

Differences from Other Canids

Compared to equally sized wolves, dogs tend to have 20 per cent smaller skulls and 10 per cent smaller brains, as well as proportionately smaller teeth than other canid species. Dogs require less calories to function than wolves. Their diet of human refuse in antiquity made the large brains and jaw muscles needed for hunting unnecessary. It is thought by certain experts that the dog's limp ears are a result of atrophy of the jaw muscles.

Dogs tend to be poorer than wolves and coyotes at observational learning, being more responsive to instrumental conditioning. Dogs have thicker skins than similairly sized wolves. Inuits tend to favour dog skin over wolf skin due to the formers resistance to tear. Unlike wolves, but like coyotes, domestic dogs have sweat glands on their pawpads. The paws of a dog are half the size of those of a wolf, and their tails tend to curl upwards, another trait not found in wolves.

Feral dogs show little of the complex social structure or dominance heirarchy present in wolf packs. For dogs, other members of their kind are of no help in locating food items, and are more like competitors.

Dogs differ from wolves and most other large canid species by the fact that they do not regurgitate food for their young, nor the young of other dogs in the same territory. Many dogs, such as this American Water Spaniel, have had their natural hunting instincts suppressed or altered to suit human needs. Like most mammals, dogs are dichromats and have colour vision equivalent to red-green colour blindness in humans.

Different breeds of dogs have different eye shapes and dimensions, and they also have different retina configurations. Dogs with long noses have a "*visual streak*" which runs across the width of the retina and gives them a very wide field of excellent vision, while those with short noses have an "*area centralis*" — a central patch with up to three times the density of nerve endings as the *visual streak* — giving them detailed sight much more like a human's.

Some breeds, particularly the sighthounds, have a field of vision up to 270° (compared to 180° for humans), although broad-headed breeds with short noses have a much narrower field of vision, as low as 180°.

Dogs detect sounds as low as the 16 to 20 Hz frequency range (compared to 20 to 70 Hz for humans) and above 45 kHz (compared to 13 to 20 kHz for humans), and in addition have a degree of ear mobility that helps them to rapidly pinpoint the exact location of a sound.

Eighteen or more muscles can tilt, rotate and raise or lower a dog's ear. Additionally, a dog can identify a sound's location much faster than a human can, as well as hear sounds up to four times the distance that humans are able to. Those with more natural ear shapes, like those of wild canids like the fox, generally hear better than those with the floppier ears of many domesticated species. Dogs have nearly 220 million smell-sensitive cells over an area about the size of a pocket handkerchief (compared to 5 million over an area the size of a postage stamp for humans).

Some breeds have been selectively bred for excellence in detecting scents, even compared to their canine brethren. What information a dog actually detects when he is scenting is not perfectly understood; although once a matter of debate, it now seems to be well established that dogs can distinguish two different types of scents when trailing, an air scent from some person or thing that has recently passed by, as well as a ground scent that remains detectable for a much longer period.

The characteristics and behaviour of these two types of scent trail would seem, after some thought, to be quite different, the air scent being intermittent but perhaps less obscured by competing scents, whereas the ground scent would be relatively permanent with respect to careful and repetitive search by the dog, but would seem to be much more contaminated with other scents.

In any event, it is established by those who train tracking dogs that it is impossible to teach the dog how to track any better than it does naturally; the object instead is to motivate it properly, and teach it to maintain focus on a single track and ignore any others that might otherwise seem of greater interest to an untrained dog.

An intensive search for a scent, for instance searching a ship for contraband, can actually be very fatiguing for a dog, and the dog must be motivated to continue this hard work for a long period of time. Domestic dogs often display the remnants of counter-shading, a common natural camouflage pattern. The general theory of countershading is that an animal that is lit from above will appear lighter on its upper half and darker on its lower half where it will usually be in its own shade.

This is a pattern that predators can learn to watch for. A countershaded animal will have dark colouring on its upper surfaces and light colouring below. This reduces the general visibility of the animal. One reminder of this pattern is that many breeds will have the occasional "blaze", stripe, or "star" of white fur on their chest or undersides. Dogs can generate large amounts of energy for a short period of time. A dog's heart and lungs are oversized relative to its body and its normal everyday needs.

A dog also has relatively more red blood cells than a human. Most of the time the dog will keep the extra red blood cells stored in its spleen. When the animal enters into a situation where its full metabolism is required, such as play, catching game, or fighting other dogs, the extra cells are released into the bloodstream.

The "oversized" heart and lungs will now be running at full capacity, and the animal will have an enhanced ability to engage in aerobic activity. This activity will produce internal heating. Dogs, being covered in fur, are limited in their ability to cool down. After a short time the animal must either cease its athletic activity or risk harming itself from overheating. One can easily observe this pattern of intense activity followed by rest periods in puppies.

During the rest phase the spleen collects red blood cells and the animal may pant to cool down. Dogs are very social animals, but their personality and behaviour vary with breed as well as how they are treated by their owners and others who come in contact with them.

It is not uncommon for dogs to attack humans and other animals; however, this is usually because of lack of care or improper upbringing by its owner. Many dogs can be trained to skillfully perform tasks not natural to canines, such as in this dog agility competition. Dogs are valued for their intelligence.

This intelligence is expressed differently with different breeds and individuals, however. For example, Border Collies are noted for their ability to learn commands, while other breeds may not be so motivated towards obedience, but instead show their cleverness in devising ways to steal food or escape from a yard.

Being highly adaptable animals themselves, dogs have learned to do many jobs as required by humans over the generations. Dogs are employed in various roles across the globe, proving invaluable assets in areas such as search-and-rescue; law enforcement (including attack dogs, sniffer dogs and tracking dogs); guards for livestock, people or property; herding; Arctic exploration sled-pullers; guiding the blind and acting as a pair of ears for the deaf; assisting with hunting, and a great many other roles which they may be trained to assume. Most dogs rarely have to deal with complex tasks and are unlikely to learn relatively complicated activities (such as opening doors) unaided.

Some dogs (such as guide dogs for the visually impaired) are specially trained to recognize and avoid dangerous situations. The meaning of "intelligence" in general, not only in reference to dogs, is hard to define. Some tests measure problem-solving abilities and others test the ability to learn in comparison to others of the same age. Defining it for dogs is just as difficult. It is likely that dogs do not have the ability to premeditate an action to solve a problem. For example, the ability to learn quickly could be a sign of intelligence. Conversely it could be interpreted as a sign of a desire to please.

In contrast, some dogs who do not learn very quickly may have other talents. An example is breeds that are not particularly interested in pleasing their owners, such as Siberian Huskies. Huskies are often fascinated with the myriad of possibilities for escaping from yards, catching small animals, and often figuring out on their own numerous inventive ways of doing both. Assistance dogs are also required to be obedient at all times. This means they must learn a tremendous number of commands, understand how to act in a large variety of situations, and recognize threats to their human companion, some of which they might never before have encountered.

Many owners of livestock guardian breeds believe that breeds like the Great Pyrenees or the Kuvasz are not easily trained because their stubborn nature prevents them from seeing the point of such commands as "sit" or "down". Hounds may also suffer from this type of ranking.

These dogs are bred to have more of a "pack" mentality with other dogs and less reliance on a master's direct commands. While they may not have the same kind of intelligence as a Border Collie, they were not bred to learn and obey commands quickly, but to think for themselves while trailing game. Dogs are highly social animals.

This can account for their trainability, playfulness, and ability to fit into human households and social situations. These attributes have earned dogs a unique position in the realm of interspecies relationships despite being one of the most effective, voracious, and potentially dangerous predators. Dogs and humans at times co-operate in some of the most effective hunting in the animal world; in that context, dogs are superpredators. The loyalty and devotion that dogs demonstrate as part of their natural instincts as pack animals closely mimics the human idea of love and friendship, leading many dog owners to view their pets as full-fledged family members.

Conversely, dogs seem to view their human companions as members of their pack, and make few, if any, distinctions between their owners and fellow dogs. Dogs fill a variety of roles in human society and are often trained as working dogs. For dogs that do not have traditional jobs, a wide range of dog sports provide the opportunity to exhibit their natural skills. In many countries, the most common and perhaps most important role of dogs is as companions.

Dogs have lived and worked with humans in so many roles that their loyalty has earned them the unique sobriquet "man's best friend". However, some cultures consider dogs to be unclean. In some parts of the world, dogs are raised as livestock to produce dog meat for human consumption.

In many places, consumption of dog meat is discouraged by social convention or cultural taboo. Laughter might not be confined or unique to humans, despite Aristotle's observation that "only the human animal laughs". The differences between chimpanzee and human laughter may be the result of the adaptations that enable human speech.

However, some behavioural psychologists argue that self-awareness of one's situation, or the ability to identify with somebody else's predicament, are prerequisites for laughter, so animals are not really laughing in the same way that humans do. The dog-laugh sounds similar to a normal pant. However by analyzing the pant using a spectrograph, this pant varies with bursts of frequencies, resulting in a laugh.

When this recorded dog-laugh vocalization is played to dogs in a shelter setting, it can initiate play, promote pro-social behaviour, and decrease stress levels. In a study by Simonet, Versteeg, and Storie, one hundred and twenty subject dogs residing in a mid-size county animal shelter were observed. Dogs ranging from 4 months to 10 years of age were compared with and without exposure to a dog-laugh recording. The stress behaviours measured included panting, growling, salivating, pacing, barking, cowering, lunging, play-bows, sitting, orienting and lying down.

The study resulted in positive findings. Exposure to the dog laughing recording resulted in the following: significantly reduced stress behaviours, increased tail wagging and the display of a play-face when playing was initiated, and more frequent pro-social behaviour such as approaching and lip licking. This research suggests exposure to dog-laugh vocalizations can calm and possibly increase shelter adoptions.

HORSE BREEDING

Horse breeding refers to reproduction in horses, and particularly the human-directed process of selective breeding of animals, particularly purebred horses of a given breed. While feral and wild horses breed successfully without human assistance, planned matings can be used to produce specifically desired characteristics in domesticated horses. Furthermore, modern breeding management and technologies can increase the rate of conception, a healthy pregnancy, and successful foaling. The male parent of a horse, a stallion, is commonly known as the sire and the female parent, the mare, is called the dam.

Both are genetically important, as each parent provides half of the genetic makeup of the ensuing offspring, called a foal. (Contrary to popular misuse, the word "colt" refers to a young male horse only; "filly" is a young female). Though many amateur horse owners may simply breed a family mare to a local stallion in order to produce a companion animal, most professional breeders use selective breeding to produce individuals of a given phenotype, or breed.

Alternatively, a breeder could, using individuals of differing phenotypes, create a new breed with specific characteristics. Some breeders consider the quality of the sire to be more important than the quality of the dam. However, other breeders maintain that the mare is the most important parent. Because stallions can produce far more offspring than mares, a single stallion can have a greater overall impact on a breed. However, the mare may have a greater influence on an individual foal because its physical characteristics influence the developing foal in the womb and the foal also learns habits from its dam when young.

Foals may also learn the "language of intimidation and submission" from their dam, and this imprinting may affect the foal's status and rank within the herd. Many times, a mature horse will achieve status in a herd similar to that of its dam; the offspring of dominant mares become dominant themselves.

Estrus Cycle of the Mare

The estrous cycle controls when a mare is sexually receptive toward a stallion, and helps to physically prepare the mare for conception. It generally occurs during the spring and summer months, although some mares may be sexually receptive into the late fall, and is controlled by the photoperiod (length of the day), the cycle first triggered when the days begin to lengthen. The estrus cycle lasts about 19-22 days, with the average being 21 days.

As the days shorten, the mare returns to a period when she is not sexually receptive, known as anestrus. Anestrus prevents the mare from conceiving in the winter months, as that would result in her foaling during the harshest part of the year, a time when it would be most difficult for the foal to survive. This cycle contains 2 phases:

- Estrus, or Follicular, phase: 5-7 days in length, when the mare is sexually recepetive to a stallion. Estrogen is secreted by the follicle. Ovulation occurs in the final 24-48 hours of estrus.
- Diestrus, or Luteal, phase: 14-15 days in length, the mare is not sexually recepetive to the stallion. The corpus luteum secretes progesterone.

Although 16 per cent of mares have double ovulations (allowing them to twin), this does not affect the length of time of estrus or diestrus.

Effects on the Reproductive System During the Estrous Cycle

Changes in hormone levels can have great affects on the physical characteristics of the reproductive organs of the mare, thereby preparing, or preventing, her from conceiving.

- Uterus: increased levels of estrogen cause edema within the uterus, making it feel heavier, and the uterus loses its tone. This edema decreases following ovulation, and the muscular tone increases. High levels of progesterone do not cause edema within the uterus. The uterus becomes flaccid during anestrus.
- Cervix: the cervix relaxes right before estrus occurs, and the secretions of the cervix increase. High progesterone levels (during diestrus) cause the cervix to close.
- Vagina: the portion of the vagina near the cervix becomes engorged with blood right before estrus. The vagina becomes relaxed and secretions increase.

- Vulva: relaxes right before estrus begins. Becomes dry, and closes tightly, during diestrus.

Hormones Involved in the Estrus Cycle, During Foaling, and after Birth

The cycle is controlled by several hormones which regulate the estrous cycle, the mare's behaviour, and the reproductive system of the mare. The cycle begins when the increased day length causes the pineal gland to reduce the levels of melatonin, thereby allowing the hypothalamus to secrete GnRH.

- *GnRH (Gonadotropin releasing hormone):* Secreted by the hypothalamus, causes the pituitary to release of 2 gonadotrophins: LH and FSH.
- *LH (Luteinizing hormone):* Levels are highest 2 days following ovulation, then slowly decrease over 4-5 days, dipping to their lowest levels 5-16 days after ovulation. Stimulates the follicle to develop, which in turn causes an increase in estrogen. Unlike most mammals, the mare does not have an increase of LH right before ovulation.
- *FSH (Follicle-stimulating hormone):* Secreted by the pituitary, causes the ovarian follicle to develop. Levels of FSH rise slightly at the end of estrus, but have their highest peak about 10 days before the next ovulation. FSH is inhibited by inhibin, at the same time LH and estrogen levels rise, which prevents immature follicles from continuing their growth.
- *Estrogen:* Secreted by the developing follicle, it causes the pituitary gland to secrete more LH (therefore, these 2 hormones are in a positive feedback loop). Additionally, it causes behavioural changes in the mare, making her more recpetive toward the stallion, and causes physical changes in the cervix, uterus, and vagina to prepare the mare for conception. Estrogen peaks 1-2 days before ovulation, and decreases within 2 days following ovulation.
- *Inhibin:* Secreted by the developed follicle right before

ovulation, "turns off" FSH, which is no longer needed now that the follicle is larger.

- *Progesterone:* Prevents conception and decreases sexual receptibility of the mare to the stallion. Progesterone is therefore lowest during the estrus phase, and increases during diestrus. It decreases about 15 days after ovulation, when the corpus luteum begns to decrease in size.
- *Prostaglandin:* Secreted by the endrometrium 13-15 days following ovulation, causes luteolysis and prevents the corpus luteum from secreting progesterone
- *eCG (equine chorionic gonadotropin - formerly called PMS/PMSG pregnant mare serum and pregnant mare serum gonadotropin):* Chorionic gonadotropins secreted if the mare conceives. First secreted by the endometrial cups around the 36th day of gestation, peaking around day 60, and decreasing after about 120 days of gestation. Also help to stimulate the growth of the fetal gonads.
- *Prolactin:* Stimulates lactation
- *Oxytocin:* Stimulates the uterus to contract

While horses in the wild mate and foal in mid to late spring, many horses domestically bred for competitive purposes, especially horse racing and various futurities, are usually born as close to January first as possible, so as to be at an advantage in size and maturity when competing against other horses in the same age group.

When an early foal is desired, barn managers will put the mare "under lights" by keeping the barn lights on in the winter to simulate a longer day, thus bringing the mare into estrus sooner than she would in nature. Mares signal estrus and ovulation by urination in the presence of a stallion, raising the tail and revealing the vulva. A stallion, approaching with a high head, will usually nicker, nip and nudge the mare, as well as sniff her urine to determine her readiness for mating.

Once fertilized, the oocyte (egg) remains in the oviduct for approximately 5.5 more days, and then descends into the uterus. The initial single cell combination is already dividing and by the time of entry into the uterus, the egg might have already reached the blastocyst stage.

The gestation period lasts for about eleven months, or about 342 days (normal average range 320-370 days). During the early days of pregnancy, the embryo (up to about 35 days), or fetus (after 35-40 days), is mobile, moving about in the uterus until about day 16 when "fixation" occurs.

True implantation - invasion into the endometrium of any sort - does not occur until about day 35 of pregnancy with the formation of the endometrial cups, and true placentation (formation of the placenta) is not initiated until about day 40-45 and not completed until about 140 days of pregnancy. A heartbeat can be detected on day 21, and the fetus gender can be determined by day 70 of the gestation.

Halfway through gestation the fetus is the size of between a rabbit and a beagle. The most dramatic fetus development occurs in the last 3 months of pregnancy when 60 per cent of fetal growth occurs.

Care of the Pregnant Mare

Domestic mares receive specific care and nutrition to ensure that they and their foals are healthy. Mares are given vaccinations against diseases such as the Rhinopneumonitis (EHV-1) virus (which can cause abortions) as well as vaccines for other conditions that may occur in a given region of the world. Mares can be used for riding or driving during most of their pregnancy, and it's healthy for them to have exercise. But only moderate exercise, especially when they become heavy in foal.

Exercise in excessively high temperatures has been suggested as being detrimental to pregnancy maintenance during the embryonic period - it should however be noted that ambient temperatures encountered during the research were in the region of 100 degrees F and the same results may not be encountered in regions with lower ambient temperatures. Mares are generally fed more than other horses and are provided supplemental vitamins and minerals because their bodies require extra nutrition to form (and later to nurse) a foal. Many feeds designed for pregnant and lactating mares also have extra fat and protein.

Overfeeding the pregnant mare should also be avoided, as excess weight may contribute to difficulties foaling or fetal/foal related problems. A special foaling stall or shed that is large and clutter free provides the mare with a safe place to give birth. Most mares foal at night or early in the morning, and prefer to give birth alone when possible. Labour is rapid, often no more than 30 minutes, and from the time the feet of the foal appear to full delivery is often only about 15 to 20 minutes.

Once the foal is born, the mare will lick the newborn foal to clean it and help blood circulation. In a very short time, the foal will attempt to stand and get milk from its mother. A foal should stand and nurse within the first hour of life. To create a bond with her foal, the mare licks and nuzzles the foal, enabling her to distinguish hers from others.

Some mares are aggressive when protecting their foals, and may attack other horses or unfamiliar humans that come near their newborns.

After birth, a foal's navel is dipped in iodine to prevent infection, it is sometimes given an enema to help clear the meconium from its digestive tract, and the newborn is monitored to ensure that it stands and nurses without difficulty.

While most horse births happen without complications, many owners have first aid supplies prepared and a veterinarian on call in case of a birthing emergency. People who supervise foaling should also watch the mare to be sure that she passes the placenta in a timely fashion, and that it is complete with no fragments remaining in the uterus, where they could cause a serious infection.

If the placenta is not removed from the stall after it is passed, a mare will often eat it, an instinct from the wild, where blood would attract predators. Foals develop rapidly, and within a few hours a wild foal can travel with the herd. In domestic breeding, the foal & dam are usually kept separate for a while, but within a few weeks are typically pastured with the other horses.

A foal will begin to eat hay, grass & grain alongside the mare at about 4 weeks old; by 10-12 weeks the foal requires more nutrition than the mare's milk can supply. Foals are typically weaned at 4-8 months of age, although in the wild a foal may nurse for a year.

How Breeds Develop

Beyond the appearance and conformation of a specific type of horse, breeders aspire to improve physical performance abilities. This concept, known as matching "form to function," has led to the development of not only different breeds, but also families or bloodlines within breeds that are specialists for excelling at specific tasks.

For example, the Arabian horse of the desert naturally developed speed and endurance to travel long distances and survive in a harsh environment, and domestication by humans added a trainable disposition to the animal's natural abilities. In the meantime, in northern Europe, the locally adapted heavy horse with a thick, warm coat was domesticated and put to work as a farm animal that could pull a plow or wagon.

This animal was later adapted through selective breeding to create a strong but ridable animal suitable for the heavily-armored knight in warfare. Then, centuries later, when people in Europe wanted faster horses than could be produced from local horses through simple selective breeding, they imported Arabians and other oriental horses to breed as an outcross to the heavier, local animals. This led to the development of breeds such as the Thoroughbred, a horse taller than the Arabian and faster over the distances of a few miles required of a European race horse or light cavalry horse.

Another cross between oriental and European horses produced the Andalusian, a horse developed in Spain that was powerfully built, but extremely nimble and capable of the quick bursts of speed over short distances necessary for certain types of combat as well as for tasks such as bullfighting. Later, the people who settled the Americas needed a hardy horse that was capable of working with cattle. Thus, Arabians and Thoroughbreds were crossed on Spanish horses, both domesticated animals descended from those brought over by the Conquistadors, and feral horses such as the Mustangs, descended from the Spanish horse, but adapted by natural selection to the ecology and climate of the west.

These crosses ultimately produced new breeds such as the American quarter horse and the Criollo of Argentina. In modern times, these breeds themselves have since been selectively bred to further specialize at certain tasks. One example of this is the American quarter horse. Once a general-purpose working ranch horse, different bloodlines now specialize in different events.

For example, larger, heavier animals with a very steady attitude are bred to give competitors an advantage in events such as team roping, where a horse has to start and stop quickly, but also must calmly hold a full-grown steer at the end of a rope. On the other hand, for an event known as cutting, where the horse must separate a cow from a herd and prevent it from rejoining the group, the best horses are smaller, quick, alert, athletic and highly trainable.

They must learn quickly, have conformation that allows quick stops and fast, low turns, and the best competitors have a certain amount of independent mental ability to anticipate and counter the movement of a cow, popularly known as "cow sense." Another example is the Thoroughbred. While most representatives of this breed are bred for horse racing, there are also specialized bloodlines suitable as show hunters or show jumpers. The hunter must have a tall, smooth build that allows it to trot and canter smoothly and efficiently.

Instead of speed, value is placed on appearance and upon giving the equestrian a comfortable ride, with natural jumping ability that shows bascule and good form. A show jumper, however, is bred less for overall form and more for power over tall fences, along with speed, scope, and agility. This favours a horse with a good galloping stride, powerful hindquarters that can change speed or direction easily, plus a good shoulder angle and length of neck. A jumper has a more powerful build than either the hunter or the racehorse.

The history of horse breeding goes back millennia. Though the precise date is in dispute, humans could have domesticated the horse as far back as approximately 4500 BCE. However, evidence of planned breeding has a more blurry history. One of the earliest people known to document the breedings of their horses were the Bedouin of the Middle East, the breeders of the Arabian horse. While it is difficult to determine how far back the Bedouin passed on pedigree information via an oral tradition, there were written pedigrees of Arabian horses by A.D. 1330. The Akhal-Teke of West-Central Asia is another breed with roots in ancient times that was also bred specifically for war and racing.

The nomads of the Mongolian steppes bred horses for several thousand years as well. The types of horses bred varied with culture and with the times. The uses to which a horse was put also determined its qualities, including smooth amblers for riding, fast horses for carrying messengers, heavy horses for plowing and pulling heavy wagons, ponies for hauling cars of ore from mines, packhorses, carriage horses and many others.

Medieval Europe bred large horses specifically for war, called destriers. These horses were the ancestors of the great heavy horses of today, and their size was preferred not simply because of the weight of the armor, but also because a large horse provided more power for the knight's lance.

Weighing almost twice as much as a normal riding horse, the destrier was a powerful weapon in battle. On the other hand, during this same time, lighter horses were bred in northern Africa and the Middle East by Muslim warriors, who preferred a faster, more agile horse.

The lighter horse suited the raids and battles of the Bedouins, allowing them to outmaneuver rather than overpower the enemy. When Muslim warriors and European knights collided in warfare, the heavy knights were frequently outmaneuvered. The Europeans, however, soon made up for the lack of speed of their native breeds by incorporating genetic traits from captured oriental horses such as the Arabian, Barb to their stables.

This cross-breeding led both to a nimbler war horse, such as today's Percheron, but also to created a type of horse known as a Courser, a predecessor to the Thoroughbred, which was used as a message horse.

During the Renaissance, horses were bred not only for war, but for haute ecole riding, derived from the most athletic movements required of a war horse, and popular among the elite nobility of the time.

Breeds such as the Lipizzan were developed from Spanish-bred horses for this purpose, and also became the preferred mounts of cavalry officers, who were derived mostly from the ranks of the nobility.

It was during this time that gunpowder was developed, and so the light cavalry horse, a faster and quicker war horse, was bred for a "shoot and run" tactic rather than the close hand-to-hand fighting seen in the Middle Ages. After Charles II retook the British throne in 1660, horse racing, which had been banned by Cromwell, was revived.

The Thoroughbred was developed 40 years later, bred to be the ultimate racehorse, through the lines of 3 foundation Arabian stallions. The 17th and 18th centuries saw more of a need for fine carriage horses in Europe, bringing in the dawn of the warmblood.

The warmblood breeds have been exceptionally good at adapting to changing times, and from their carriage horse beginnings they easily transitioned during the 1900s into a sport horse type. Today's warmblood breeds, although still used for competitive driving, are more often seen competing in the show jumping or dressage arenas.

The Thoroughbred continues to dominate the horseracing world, although its lines have been more recently used to improve warmblood breeds and to develop sport horses. The predecessor of the American Quarter Horse was developed in the 1700s, mainly for quarter racing (racing ¼ of a mile). The breed was later adapted for work in the west, and "cow sense" was particularly bred for as their use for herding cattle increased. However, because there was also a need for animals suitable for sprint racing, the modern Quarter Horse has two distinct types: the sleeker racing type and the stock horse type.

The racing type most resembles the finer-boned ancestors of the first racing Quarter Horses, and the type is still used for ¼-mile races.

The stock horse type, used in western events, is bred for a shorter stride, docile temperament, and cow sense. The need for horses for heavy draft and carriage work continued until the industrial revolution and the advent of the automobile and the tractor.

After this time, draft and carriage horse numbers dropped significantly, though light riding horses remained popular for recreational pursuits. Draft horses today are used on a few small farms, but today are seen mainly for pulling and plowing competitions rather than farm work.

Heavy harness horses are now used as an outcross with lighter breeds, such as the Thoroughbred, to produce the modern warmblood breeds popular in Olympic and sport horse disciplines. Breeding a horse is an endeavor where the owner, particularly of the mare, will usually need to invest considerable time and money.

For this reason, a horse owner needs to consider several factors, including:

- Does the proposed breeding animal have valuable genetic qualities to pass on?
- Is the proposed breeding animal in good physical health, fertile, and able to withstand the vigors of reproduction?
- For what purpose will the foal be used?
- Is there a market for the foal in the event that the owner does not wish to keep the foal for its entire life?
- What is the anticipated economic benefit, if any, to the owner of the ensuing foal?
- What is the anticipated economic benefit, if any, to the owner(s) of the sire and dam or the foal?
- Does the owner of the mare have the expertise to

properly manage the mare through gestation and parturition?

- Does the owner of the potential foal have the expertise to properly manage and train a young animal once it is born?

There are value judgements involved in considering whether an animal is suitable breeding stock, hotly debated by breeders. Additional personal beliefs may come into play when considering a suitable level of care for the mare and ensuing foal, the potential market or use for the foal, and other tangible and intangible benefits to the owner. If the breeding endeavor is intended to make a profit, there are additional market factors to consider, which may vary considerably from year to year, from breed to breed, and by region of the world.

In many cases, the low end of the market is saturated with horses, and the law of supply and demand thus allows little or no profit to be made from breeding unregistered animals or animals of poor quality, even if registered.

The minimum cost of breeding for a mare owner includes the stud fee, and the cost of proper nutrition, management and veterinary care of the mare throughout gestation, parturition, and care of both mare and foal up to the time of weaning. Veterinary expenses may be higher if specialized reproductive technologies are used or health complications occur. Making a profit in horse breeding is often difficult.

While some owners of only a few horses may keep a foal for purely personal enjoyment, many individuals breed horses in hopes of making some money in the process. A general rule of thumb is that a foal intended for sale should be worth three times the cost of the stud fee if it were sold at the moment of birth. From birth forward, the costs of care and training are added to the value of the foal, with a sale price going up accordingly. If the foal wins awards in some form of competition, that may also enhance the price.

On the other hand, without careful thought, foals bred without a potential market for them may wind up being sold at a loss, and in a worst-case scenario, sold for "salvage" value — a euphemism for sale to slaughter as horsemeat. Therefore, a mare owner must consider their reasons for breeding, asking hard questions of themselves as to whether their motivations are based on either emotion or profit and how realistic those motivations may be.The stallion should be chosen to complement the mare, with the goal of producing a foal that has the best qualities of both animals, yet avoids having the weaker qualities of either parent.

Generally, the stallion should have proven himself in the discipline or sport the mare owner wishes for the "career" of the ensuing foal. Mares should also have a competition record showing that they also have suitable traits, though this does not happen as often. A purebred horse is usually worth more than a horse of mixed breeding, though this matters more in some disciplines than others. The breed of the horse is sometimes secondary when breeding for a sport horse, but some disciplines may prefer a certain breed or a specific phenotype of horse.

Sometimes, purebred bloodlines are an absolute requirement: For example most Racehorses in the United States must be recorded with a breed registry in order to race.Bloodlines are often considered, as some bloodlines are known to cross well with others.

If the parents have not yet proven themselves by competition or by producing quality offspring, the bloodlines of the horse are often a good indicator of quality and possible strengths and weaknesses. Some bloodlines are known not only for their athletic ability, but could also carry a conformational or genetic defect, poor temperament, or for a medical problem.

Some bloodlines are also fashionable or otherwise marketable, which is an important consideration should the mare owner wish to sell the foal. Horse breeders also consider conformation, size and temperament. All of these traits are heritable, and will determine if the foal will be a success in its chosen discipline.

The offspring, or "get," of a stallion are often excellent indicators of his ability to pass on his characteristics, and the particular traits he actually passes on. Some stallions are fantastic performers but never produce offspring of comparable quality. Others sire fillies of great abilities but not colts. At times, a horse of mediocre ability sires foals of outstanding quality. Mare owners also look into the question of if the stallion is fertile and has successfully "settled" (i.e. impregnated) mares. A stallion may not be able to breed naturally, or old age may decrease his performance. Mare care boarding fees and semen collection fees can be a major cost.

Breeding a horse can be an expensive endeavor, whether breeding a backyard competition horse or the next Olympic medalist. Costs may include:

- The stud and booking fee
- Fees for collecting, handling, and transporting semen (if AI is used and semen is shipped)
- Mare exams: to determine if she is healthy enough to breed, to determine when she ovulates, and (if AI is used) to inseminate her
- Mare transport, care, and board if the mare is bred live cover at the stallion's residence
- Veterinary bills to keep the pregnant mare healthy while in foal
- Possible veterinary bills during pregnancy or foaling should something go wrong
- Veterinary bills for the foal for its first exam a few days following foaling

Stud fees are determined by the quality of the stallion, his performance record, the performance record of his get (offspring), as well as the sport and general market that the animal is standing for. The highest stud fees are generally for racing Thoroughbreds, which may charge from two to three thousand dollars for a breeding to a new or unproven stallion, to several hundred thousand dollars for a breeding to a stakes winner.

Sport horse stallions generally range from \$1000 to \$3000, although the top stallions may reach \$4000 for one breeding. The lowest stud fees may only be \$100-\$200, but there are trade-offs: the horse will probably be unproven, and probably much less athletic than a horse with a stud fee that is a bit higher. As a stallion's career, either performance or breeding, improves, his stud fee tends to increase in proportion. If one or two offspring are especially successful, winning several stakes races or an Olympic medal, the stud fee will generally greatly increase.

Younger, unproven stallions will generally have a lower stud fee earlier on in their careers. To help decrease the risk of financial loss should the mare die or abort the foal while pregnant, many studs have a live foal guarantee (LFG), allowing the owner to have a free breeding to their stallion the next year. However, this is not offered for every breeding.

Covering the Mare

After the mare is bred or artificially inseminated, she is checked 16 days later to see if she "took", and is pregnant. A second check is usually performed at 28 days. If the mare is not pregnant, she may be bred again during her next cycle. It is considered safe in the horse to breed a mare to a stallion of much larger size.

Because of the mare's type of placenta and its attachment and blood supply, the foal will be limited in its growth within the uterus to the size of the mare's uterus, but will grow to its genetic potential after it is born. Test breedings have been done with draft horse stallions bred to small mares with no increase in the number of difficult births.

When breeding live cover, the mare is usually boarded at the stud. She is "teased" several times with a stallion that will not breed to her, usually with the stallion being presented to the mare over a barrier. Her reaction to the teaser, whether hostile or passive, is noted. A mare that is in heat will generally tolerate a teaser (although this is not always the case), and may present herself to him, holding her tail to the side.

A veterinarian may also determine if the mare is ready to be bred, by ultrasound or palpating daily to determine if ovulation has occurred. When it has been determined that the mare is ready, both the mare and intended stud will be cleaned. The mare will then be presented to the stallion, usually with one handler controlling the mare and one or more handlers in charge of the stallion. Multiple handlers are preferred, as the mare and stallion can be easily separated should there be any trouble. The Thoroughbred industry requires all registered foals to be bred through live cover. Artificial fertility treatments, listed below, are not permitted.

By contrast, the standardbred industry allows registered foals to be bred by live cover, or by artificial insemination (AI) with fresh or frozen (not dried) semen. No other artificial fertility treatment is allowed. In addition, foals bred via AI may only be registered if the stallion's sperm was collected during his lifetime, and used no later than the calendar year of his death or castration.

Artificial Insemination

Artificial insemination (AI) has several advantages over live cover, and has a very similar conception rate:

- The mare and stallion never have to come in contact with each other, which therefore reduces breeding accidents, such as the mare kicking the stallion.
- AI opens up the world to international breeding, as semen may be shipped across continents to mares that would otherwise be unable to breed to a particular stallion.
- A mare also does not have to travel to the stallion, so the process is less stressful on her, and if she already has a foal, the foal does not have to travel.
- AI allows more mares to be bred from one stallion, as the ejaculate may be split between mares.
- AI reduces the chance of spreading sexually transmitted diseases between mare and stallion.
- AI allows mares or stallions with health issues, such as sore hocks which may prevent a stallion from mounting, to continue to breed.
- Frozen semen may be stored and used to breed mares even after the stallion is dead, allowing his lines to continue. However, the semen of some stallions does not freeze well, and live cover is sometimes the only option. Also, as noted earlier with respect to

standardbreds, a particular breed registry may ban the use of frozen semen in posthumous breeding.

A stallion is usually trained to mount a phantom mare, although a live mare may be used, and he is collected using an artificial vagina (AV), which is often heated to simulate the vagina of the mare. The AV has a filter and collection area at one end to collect the semen, which is then processed in a lab. The semen is then chilled or frozen and shipped to the mare owner.

When the mare is in heat, a veterinarian introduces the semen directly into her via a syringe and pipette. Often an owner does not want to take a valuable competition mare out of training to carry a foal. This presents a problem, as the mare will usually be quite old by the time she is retired from her competitive career, at which time it is more difficult to impregnate her. Other times, a mare may have physical problems that prevent or discourage breeding. However, there are now several options for breeding these mares. These options also allow a mare to produce multiple foals each breeding season, instead of the usual one. Therefore, mares may have an even greater value for breeding.

Embryo Transfer

The relatively new method involves removing the mare's fertilized embryo a few days following insemination, and transferred to a surrogate mare.

Gamete Intrafallopian Transfer (GIFT)

The mare's ovum and the stallion's sperm are deposited in the oviduct of a surrogate dam. This technique is very useful for subfertile stallions, as fewer sperm are needed, so a stallion with a low sperm count can still successfully breed.

Egg Transfer

An oocyte is removed from the mare's follicle and transferred into the oviduct of the recipient mare, who is then bred. This is best for mares with physical problems, such as an obstructed oviduct, that prevent breeding.

COW BREEDING

Different Breeds for Different Needs

Dairy cattle are breeds that have been developed to produce milk whereas beef cattle are bred and selected primarily for the production of meat. Dual-purpose breeds do exist and they have been selected for both meat and milk production. In the last 20 years the genetic make up of the UK dairy herd has changed and is still changing.

Cows are larger, earlier maturing, are producing far more milk and are more efficient at converting feed into milk. As it is the birth of the calf that causes milk production, it is critical for the farmer that the cow becomes pregnant again during lactation and produces the next calf one year after the previous one. The average length of productive life of a dairy cow is between four and five lactations, although this varies from one or two to over ten. Since the fifties artificial insemination (AI) has increasingly replaced natural service. Embryo transplants (ET) are also being developed.

A surrogate mother may have difficulty delivering, since the foreign calf is often larger than nature intended. When this happens, a rope or chain is attached to the calf's forelegs, and, with a winch-like device the newborn calf is pulled from the womb. Unnecessary suffering for both cow and calf.

Regular Calving

A high annual milk production per cow is dependent on regular calving at about 12- month intervals. Poor reproductive efficiency leads to longer calving indices and a reduction in milk sales per cow per annum. It also leads to an increase in the number of cows culled for failing to conceive. On average, in the UK, only about 85 cows calve per annum for every 100 cows available for breeding. Over thirty-five per cent of all cows culled are disposed of because of reproductive disorders.

Obtaining Semen

Photographer Henry Heap visited the Milk Marketing Board stud bull operation in Mid Wales. He provided Turning

Point magazine with a complete set of photographs and commentary of his visit. "A large, handsome, tan bull stood stock still, tethered by the ring tethered through his nose to what looked like a giant circular clothes line or some grotesque merry-go-round. He was kept so motionless by a rubber hood that covered his eyes, which had a similar effect to that of hooding a falcon.

A worker commented that the bull would stay still all day if they left the blindfold on. However, this bull was not the source of the semen, but instead what is known as a 'teaser'. The stud bulls are led out by the nose one at a time to mount the motionless hooded bull. After they have mounted the teaser their penis is diverted into a leather vagina which is used to collect the semen.

As soon as the process is complete, another bull is led out. The hood keeps the teaser bull stationery for each successive beast to mount. These stud bulls never even see females. During embryo transfer an embryo can be flushed from the uterus of a genetically superior cow and implanted in a receiver cow of so-called 'poorer quality' for foetal development.

Embryos are also taken from dead cows at the slaughterhouse. Hormone treatment is used to induce multiple ovulation and the cow is then inseminated. A few days later several embryos can be flushed from her. For both embryo collection and implanting the cow receives a local anaesthetic injection in the tail. Embryos of the desired breed can then be used. ET maximises the number of offspring from selected breeding females and maximises profit. Do-it-yourself embryo transfer is now taking place in the UK. Courses are run to train in direct embryo transfer techniques using the quick-thaw method for frozen embryos.

Although vets are not required to be present at the time of transfer, they must check each animal is suitable for implantation. Farmers genetically improve the herd by using techniques such as MOET (multiple ovulation embryo transfer) which involves inducing a genetically 'superior' cow to produce up to 20 embryos rather than one. For ease of management, farmers prefer cows to calf at the same time. They, therefore, need to detect when cows come into oestrous. The oestrous detection rate (ODR) is the number of cows observed in oestrus during a 3 week period, in relation to the number of cows expected to be observed.

The average ODR on farms is about 60 per cent. One alternative is to synchronise the oestrous period of several animals by injections of prostaglandin or administration of progesterone as an impregnated intraveginal coil implanted for 9-12 days.

Artificial Insemination

Artificial insemination (AI) now accounts for around 75 per cent of British cattle births. The cow is tied up and one hand of the inseminator manipulates the cervix through the rectum wall whilst the other discharges semen into the vagina and cervix using an inseminating gun. A regional manager with AI firm Genus the problems associated with the procedure. "Placing a quarter of a cc of diluted semen into an area as small as a penny coin, when the target area is inside 500kg of unco-operative cow, requires a high degree of expertise.

The more the cow is able to move about, the harder it becomes to hit the target. Poor handling facilities cause stress to you, the cow, the inseminator and both our bank balances." Natural service by bulls is still used, but the risk of slowing down genetic progress, and the danger of handling bulls means farmers mainly use AI. There are many invasive procedures carried out on farm animals that the general public rarely witness. One of many includes making sure the cow is in calf. This is carried out using an ultrasound device much like a small tv set with a long thick wire attachment.

A vet will insert their hand and arms length into the vagina of the cow. The animal is held in a holding device to keep her still. The screen shows the small calf in the middle of a dark shaped uterus. Herds usually calve at the same time (due to human intervention) so all the cows on the farm will usually be checked at the same time.

Jersey Cattle

A small, honey-brown breed of dairy cattle, the Jersey is renowned for the high butterfat content of its milk, as well as a genial disposition.

The Jersey cow is quite small, ranging from only 800 to 1200 pounds (360 to 540 kg). The main factor contributing to the popularity of the breed has been their greater economy of production, due to:

- The ability to carry a larger number of effective milking cows per unit area due to lower body weight, hence lower maintenance requirements
- High butterfat conditions, and to thrive on locally produced food.

Bulls are also small, ranging from 1200 to 1800 pounds (540 to 820 kg), and are notoriously very aggressive. Castrated males can be trained into fine oxen which, due to their small size and gentle nature make them popular with young teamsters. Jersey oxen are not as strong as larger breeds however and are generally out of favour among competitive teamsters.

Due to the small size, curious docile character and attractive features of the Jersey cow, small herds were imported into England by aristocratic landowners as props for aesthetic landscapes. As its name implies, the Jersey was bred on the British Channel Island of Jersey.

It apparently descended from cattle stock brought over from the nearby Norman mainland, and was first recorded as a separate breed around 1700. Frank Falle has speculated (on the basis of DNA evidence) that the oldest settlers in Jersey were Danish vikings who had been to Nantes with Hatain, where it is recorded that tribute was given by the King of France to Hatain of 500 cattle to leave that area, whereupon they settled in the Normandy area.

The marked resemblance between Jersey cattle and the Nantaise cow would seem to bear this out.

While the breed is isolated from outside influence today, this was not always the case. Before 1789 cows would be given as dowry for inter-island marriages between Jersey and

Guernsey. This was, however, not widespread. Since 1789, imports of foreign cattle into Jersey have been forbidden by law to maintain the purity of the breed, although exports of cattle and semen have been an important economic resource for the island.

The restriction on the import of cattle was initially introduced in 1789 to prevent a collapse in the export price. The United Kingdom levied no import duty on cattle imported from Jersey. Cattle were being shipped from France to Jersey and then being shipped onward to England to circumvent the tariff on French cattle.

The increase in the supply of cattle, sometimes of inferior quality, was bringing the price down and damaging the reputation of Jersey cattle. The import ban stabilised the price and enabled a more scientifically controlled programme of breeding to be undertaken.

Sir John Le Couteur studied selective breeding and became a Fellow of the Royal Society - his work led to the establishment of the Royal Jersey Agricultural and Horticultural Society in 1833. At that time, the breed displayed greater variation than it does today with white, dark brown and mulberry beasts.

SHEEP BREEDS

The domestic sheep is a multi-purpose animal, and the more than 200 breeds now in existence were created to serve these diverse purposes. Some sources give a count of a thousand or more breeds, but concrete documentation has not been produced for such a number. Almost all sheep are classified as being best suited to furnishing a certain product: wool, meat, milk, hides or a combination thereof (i.e. dual-purpose breeds). Other features that are important when classifying sheep include face colour (generally white or black), the length of the tail, the presence or lack of horns, and the topography for which the breed has been developed.

This last point is especially stressed in the United Kingdom, where breeds are described as either upland (hill or mountain) or lowland breeds. A sheep may also be of a fat-tailed breed, which is a type of dual-purpose sheep common in Africa and Asia with larger deposits of fat within its tail, or the meat from such a breed. Ovine breeds are also grouped based on how well they are suited to producing a certain type of breeding stock. Generally, sheep are thought to be either "ewe breeds" or "ram breeds". Ewe breeds are those that are hardy, and have good reproductive and mothering capabilities—they are for replacing breeding ewes in standing flocks.

Ram breeds are selected for rate of weight gain and carcass quality, and are mated with ewe breeds to produce meat lambs. Many breeds, especially rare or primitive ones, fall into neither category strictly. Lowland and upland breeds are also crossed in this fashion, with the hardy hill ewes crossed with larger, fast-growing lowland rams to produce a type of ewe called a mule, which can then be crossed with meat-type rams to produce prime market lambs.

Sheep breeds are categorized by the type of wool they exhibit, irrespective of whether they are favoured for meat or wool production. Fine wool breeds are those that have wool of great crimp and density, and are consequently preferred by the textile industry. Most of these breeds were derived from Merino sheep, and Merinos continue to dominate the world sheep industry.

The record for the most valuable sheep ever belongs to an Australian Merino ram which was sold for $16,000 AUD. Medium wool breeds have wool between the extremes, and are typically fast-growing meat and ram breeds with dark faces. Some major medium wool breeds, such as the Corriedale, are dual-purpose crosses of long and fine-wooled breeds created for high-production commercial flocks. Long wool breeds are the largest of sheep, with lengthy wool and a slow rate of maturation.

Long wool sheep are most valued by modern agriculture for crossbreeding to improve the attributes of other sheep types.

For example: the American Columbia breed was developed through crossing Lincoln rams—a long wool breed—with fine-wooled Rambouillet ewes. Coarse or carpet wool sheep are those with a medium to long length wool of characteristic coarseness. Breeds traditionally used for carpet wool show more variability in general traits than other types, the chief requirement being a wool that will not break down under heavy use (as would that of the finer breeds).

As the demand for carpet-quality wool declines, some breeders of this type of sheep are attempting to reposition a few of the traditional breeds to alternative purposes. Others have always been primarily meat-class sheep. A last group of sheep breeds is that of fur or hair sheep, which do not grow wool at all. Hair sheep are similar to the very early type of domesticated sheep kept before wooly breeds were developed, and are primarily raised for meat and pelts. Some modern breeds of hair sheep, such as the Dorper, are the result of crosses between wooled and hair breeds.

For producers uninterested in selling wool, hair sheep are highly advantageous, as those who keep hair sheep need not go to the expense of shearing. Hair sheep are also resistant to parasites and hot weather, and are consequently increasing in popularity among a subset of producers. With the modern rise of corporate agrobusiness and the decline of localized family farms, many breeds of sheep are in danger of extinction.

The Rare Breeds Survival Trust of the U.K. lists 25 native breeds as having only 3,000 registered animals, and the American Livestock Breeds Conservancy lists 14 as being at less than 10,000.

Preference for only those breeds with uniform characteristics and fast growth rates has pushed heritage (or heirloom) breeds to the margins of the sheep industry. Those that remain are maintained through the efforts of conservation organizations, breed registries, and individual farmers dedicated to their preservation.

Chapter 13

Veterinary Science

Veterinary science is the study, diagnosis, treatment and prevention of disease in animals both as individuals and as groups. There is also a key role for members of the profession as guardians of human health in the context of disease transmission from animal or animal products to man.

Once early man had moved from hunting to herding he had an interest in the health and husbandry of his stock. By the time of the ancient civilisations there is evidence of close interaction between man and animals in contexts which are still familiar today including the use of animals for production, draught and war and, increasingly, as human companions. The first book of the modern era devoted to veterinary medicine, 'Artis Veterinariae', was produced by Publius Vegetius Renatus in the second half of the 5th century.

The 'veterinarii' were the animal doctors of ancient Rome, and the word came back into use in the 17th and 18th centuries as the veterinary profession emerged from its origins amongst the farriers. By the 18th century several veterinary texts had been published and the first veterinary school in Europe was established at Lyon in 1762. The first to be established in the United Kingdom was the Royal Veterinary College in 1792.

Although the term 'veterinary surgeon' is widely used within the UK, the term 'veterinarian' is employed here. The veterinary workplace has changed in the last century with less emphasis on the horse (especially as a draught animal), an increasing emphasis on companion animals kept for pleasure, and greater veterinary involvement in production animals, public health and food hygiene.

The role of the profession in protecting the health and welfare of more diverse species groups such as laboratory animals, zoological collections, wildlife and, indeed, the contribution to conservation of endangered species, continues to grow.

There is also an increasing body of specialists, both in individual species and in disciplines which cut across species groups, as opportunities for training and the demand for specialist services have grown. The comparative approach of veterinary science will continue to provide insight and support for basic scientists and contribute to the understanding of human disease.

The beginning of the 21st century finds the veterinary profession and its work held in high esteem by the general public and a source of considerable interest, with unprecedented exposure of veterinary matters in the popular media. Veterinarians are regarded as guardians of animal health and welfare, and the veterinary schools have a responsibility to continue to produce graduates in whom the public will have confidence.

The sustained public appeal of veterinary work has led to a level of demand for places on veterinary courses which far exceed supply. This demand enables the schools to select strongly motivated, high achievers with entry qualifications among the highest in UK university courses. Most applicants are attracted in the first instance by the prospect of veterinary clinical practice with its unique combination of science, art, practical skills, human-animal and interpersonal interaction. However, an increasing number follow other career paths as they become aware of the diverse opportunities provided by a veterinary degree.

Challenges to Today's Veterinary Medicine

Today, the veterinary profession is confronted with a different set of problems compared with the middle of last century. Now, veterinarians often have to deal with herds or regions remaining diseased after lengthy disease control campaigns. In addition, it is considered necessary to take into account the economic aspects of disease control through the use of benefit/cost analyses of disease control campaigns. Costly multi-factorial *disease complexes* such as mastitis or cattle lameness have become quite common. In addition, old, emerging or new *diseases* with complex aetiologies pose a difficult challenge for the profession. Veterinarians have to respond to the challenges posed by problems such as BSE, FMD and antimicrobial resistance. All these problems require identification, quantification and intensive examination of multiple, directly or indirectly causal, and often interacting, disease determinants.

Veterinary epidemiology and evidence-based veterinary medicine provide overlapping sets of tools which can be used to approach these new challenges. In clinical practice, there is now a daily need for valid, up-to-date information about diagnosis, therapy and prevention. This situation is complicated by the inadequacy of traditional sources for this type of information, since they may be out-of-date, they are frequently wrong, it may be ineffective (didactic CPD), too overwhelming in volume, or too variable in validity. In addition, the clinician

has to deal with the disparity between diagnostic skills and clinical judgement (increasing with experience) on the one hand and up-to-date knowledge and clinical performance (declining with time) on the other. The fact that he/she has insufficient time to examine the animal patient and to practice CPD further complicates the situation.

VETERINARY EPIDEMIOLOGY

Veterinary epidemiology deals with the investigation of diseases, productivity and animal welfare in populations. It is used to describe the frequency of disease occurrence and how disease, productivity and welfare are affected by the interaction of different factors or determinants. This information is then used to manipulate such determinants in order to reduce the frequency of disease occurrence.

Veterinary epidemiology is a holistic approach aimed at co-ordinating the use of different scientific disciplines and techniques during an investigation of disease or impaired productivity or welfare. The field of veterinary epidemiology can be divided into different components.

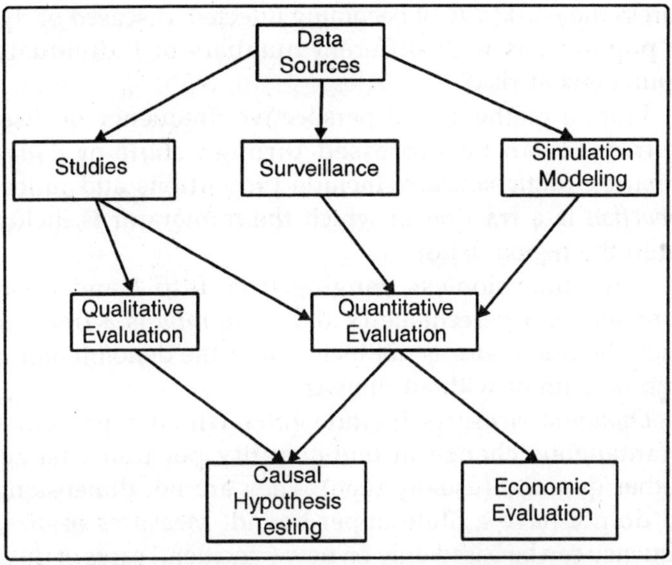

Fig. Components of Veterinary Epidemiology

One of its essential foundations is the collection of data, which then has to be analysed using qualitative or quantitative approaches in order to formulate causal hypotheses. As part of the quantitative approach to epidemiological analysis, epidemiological investigations involving field studies or surveys are being conducted and models of epidemiological problems can be developed. The ultimate goal is to control a disease problem, reduce productivity losses and improve animal welfare.

Measurement of Disease Frequency and Production

One of the most fundamental tasks in epidemiological research is the quantification of the disease occurrence. This can be done simply on the basis of *counts* of individuals which are infected, diseased, or dead. This information will be useful for estimating workload, cost, or size of facilities to provide health care. More commonly, *counts* are expressed as a *fraction* of the number of animals capable of experiencing infection, disease or death.

These types of quantities are used by epidemiologists to express the *probability* of becoming infected, diseased or dying for populations with different numbers of individuals (= populations at risk).

From a mathematical perspective, frequency of disease occurrence can be expressed through *static* or *dynamic* measures. *Static measures* include proportions and ratios. A *proportion* is a fraction in which the numerator is included within the denominator.

It is dimensionless, ranging from 0 to 1 and is often expressed as a percentage (x 100). The *ratio* is a fraction in which the numerator is not included in the denominator and it can be with or without dimension.

Dynamic measures include *rates* which represent the instantaneous change in one quantity per unit change in another quantity (usually time). They are not dimensionless and do not have a finite upper bound. Measures of disease frequency can be based only on new (=incident) cases of disease or do not differentiate between old and new disease.

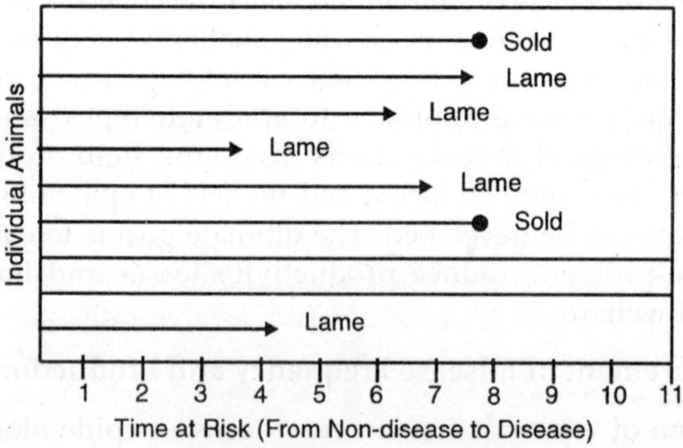

Fig. Incidence of Disease

Figure shows the principles behind incidence measures. They are derived from data for animals which did not have the disease at the beginning of the study period. These animals are followed over time until they develop the disease (eg become lame) or until the observation period finishes (eg sold or end of study period).

Cumulative Incidence

The risk of new disease occurrence is quantified using *cumulative incidence,* also called *incidence risk.* It is defined as the proportion of disease-free individuals developing a given disease over a specified time, conditional on that individual's not dying from any other disease during the period.

Note that animals have to be disease free at the beginning of the observation period to be included in the enumerator or denominator of this calculation. It is interpreted as an individual's risk of contracting disease within the risk period. The quantity is dimensionless, ranges from 0 to 1 and always requires a period referent.

As an example, last year a herd of 121 cattle were tested using the tuberculin test and all tested negative. This year, the same 121 cattle were tested again and 25 tested positive. The cumulative incidence over a period of 12 months would then be calculated as 25/121 which amounts to 0.21. Hence, an individual animal within this herd had a 21% chance of becoming infected over the 12 month period.

Miscellaneous Measures of Disease Occurrence

Other measures of disease frequency include the *attack rate* which is defined as the number of new cases divided by the initial population at risk. As it is based on the same calculation as cumulative incidence it really is a subtype of cumulative incidence. It is confusing though that despite its name it is in fact a probability and not a rate. The attack rate is used when the period at risk is short.

Mortality rates are applied using a number of different interpretations, and often do not represent a true rate. The *crude mortality rate* has death as the outcome of interest and is calculated analogous to incidence density. The *cause-specific mortality rate* is estimated for specific causes of death and also calculated analogous to incidence density. *Case fatality rate* represents the proportion of animals with a specific disease that die from it. It is a risk measure, not a rate, and is used to describe the impact of epidemics or the severity of acute disease.

Time to the occurrence of an event such as death or onset of clinical disease can be estimated for many epidemiological data sets describing repeated observations on the same sample of animals. This type of data can be summarised using for example mean survival time. This particular method has the disadvantage that the estimate will depend on the length of the time period over which data was collected. If the interval is too short, survival time is likely to be estimated incorrectly, as only individuals who experienced the event of interest can be included in the calculation.

As an alternative, it is possible to calculate an incidence density by using person-years of observation in the denominator and number of events in the enumerator. But this calculation assumes that the rate at which the event occurs is constant throughout the period of study. The most appropriate technique for this data is based on the *survivor* and *hazard function*. The survivor function (=cumulative survival probability) is a functional representation of the proportion of individuals not dying or becoming diseased beyond a given time at risk.

It can be interpreted as the probability of remaining alive for a specific length of time. The survival function is often summarized using the *median survival time* which is the time at which 50% of individuals at risk have failed (died or became diseased).

The *hazard function* (=instantaneous failure rate, force of mortality, conditional mortality rate, age specific failure rate) is calculated by dividing the conditional probability of an individual dying or becoming diseased during a specific time interval provided it has not died or become diseased prior to that time divided by the specified *time interval*. This parameter does represent a rate expressing the potential of failing at time t per unit time given survival up until time t.

In the context of survival data, *censoring* is an extremely important concept. In the case of *right censoring*, individuals are lost to follow-up or are not dead/diseased at the end of the follow-up period.

This particular type of censoring can be easily accounted for in survival analysis by excluding them from the denominators of the calculations following their departure from the population at risk. With *left censoring,* beginning of the time at risk is not known, and the commonly used analysis techniques cannot take account of this type of censoring.

An example calculation for survival data is presented in Figure Animal A survived for 4 months, animal B survived the whole period of 12 months and animal C was removed from the population after 7 months. The number of survivors is based on the actual number of animals still alive after a given time period.

The value for the cohort is used as the denominator for calculation of cumulative survival. The number is adjusted for censored observations. The failures represents the number of deaths during a particular time interval. And the hazard rate is the probability of death per unit time (one month in this case).

STANDARDISATION OF RISK

A crude risk estimate summarises the effects of the specific risk and the subgroup distribution. But in the presence of a confounding factor a crude risk estimate may distort the true pattern. In such a situation, patterns of disease should be described using host-attribute specific risk estimates (e.g. age, sex).

For each level of the attribute a separate stratum is formed and stratum-specific risk estimates are calculated. Summary figures can be produced using *standardised* or *adjusted risk* estimates. The two main methods available to perform these calculations are *direct* and *indirect standardisation.*

Direct Standardization

Direct standardization involves weighting a set of observed category specific risk estimates according to a *standard* distribution. First, stratum-specific risk rates are calculated. Then a *standard* population distribution is estimated and the proportion of the standard population in each stratum calculated.

The direct adjusted risk rate estimate is obtained as the sum of the products across the strata between the proportion of the *standard* population in stratum *i* and the observed risk rate estimate in stratum *i* in the *study* population. As an example, the mortality in humans in 1963 was compared between Sweden and Panama. In this particular year, Sweden had a population size of 7.496.000 and 73.555 deaths resulting in a *mortality rate* of 0.0098 per year.

Panama had a population size of 1.075.000 with 7871 deaths giving a *mortality rate* of 0.0073 per year. Based on these figures it appeared that life in Sweden was more risky than in Panama. It becomes apparent that in his comparison the confounding factor was the difference in age structure between the two populations.

Sweden had a much lower mortality in young people, but because it had a large proportion of old people this effect did not come through in the aggregated analysis. After adjustment for differences in age structure it turns out that mortality was higher in Panama with 0.0162 per year than it was in Sweden with 0.015 per year.

Indirect Standardisation

An alternative approach to standardisation of risk estimates is called *indirect standardisation*. In this case, a *standard* population does not supply the weighting distribution, but a set of stratum-specific risk estimates which are then weighted to the distribution of the *study* population. This technique is used if stratum-specific risk estimates are not available.

But in order to be able to use the method, stratum-specific risk estimates for the *standard* population and the frequency of the adjusting factor in the *study* population have to be available.

As a first step, the expected number of cases is calculated on the basis of the sum of the products between stratum-specific rates for the *standard* population and the total number of individuals in each stratum in the *study* population. The standardised morbidity or mortality ratio (SMR) is calculated using the number of observed cases divided by the number of expected cases, and the indirect adjusted risk is obtained from multiplying the overall observed risk in the *study* population with the SMR.

ANALYTICAL EPIDEMIOLOGY

The major aims of epidemiology are to *describe* the health status of populations, to *explain* the aetiology of diseases, to *predict* disease occurrence and to *control* the distribution of disease. An understanding of causal relationships is the basis of the last three objectives.

Such associations between causes and disease occurrence can be discovered through individual *case studies*, by experimental *laboratory studies* and by *field studies*. *Case studies* focusing on individual sick animals have long been at the centre of clinical knowledge.

They are based on direct personal observations relating to anatomical structure and physiological function, which can be quantified and are systematic but still largely qualitative. While these observations can be extremely intensive and detailed their disadvantage is their subjectivity and the possibly extreme variation between cases. In the *laboratory experiment* - the classic experiment- great precision in measurements and optimal control of influencing variables can be achieved resulting in sound inferences.

The disadvantage is that it is usually not possible to represent the myriad of factors affecting disease occurrence in the natural environment of the animal and it may be difficult to work with sufficient numbers of animals to represent true variation between animals in the natural population.

A *field study* is conducted in the natural environment of the animals and measurements are made on sick as well as healthy animals.

The differences between sick and healthy animals can be described with respect to the frequency of presence or absence of potential risk factors. With this type of study, animals are exposed to all the known and unknown environmental factors present in their natural environment.

Field research is *empirical* and involves measurement of variables, estimation of population parameters and statistical testing of hypotheses. It is of a *probabilistic nature* in that as a result of a population study it will not be possible to predict with certainty which animal will develop a particular disease given the presence of certain risk factors.

But it will be possible to predict how many cases of the disease will occur in the population in the future. Field research involves comparisons among groups in order to estimate the magnitude of an *association* between a putatively causal factor and a disease.

The objective is to assess if there is the potential of a cause-effect relationship between a single or multiple risk factors and the disease.

The interrelatedness of phenomena within a biological system complicates the situation for an investigator who will always have to select a segment of the system for the investigation.

Attempting to isolate the segment from the rest of the system can result in an outcome which does not represent the real situation in the system anymore. *Analytical epidemiology* is aimed at determining the *strength, importance* and *statistical significance* of epidemiological associations. The process typically begins with data collection and eventually leads to data analysis and interpretation.

The data collection can be based on a *survey* or a *study*. Both terms are often used interchangeably. A *survey* typically involves *counting* members of an aggregate of units and *measuring* their characteristics. In contrast, a *study* is aimed at comparison of different groups and investigation of cause-effect relationships.

Both designs can be based on a *census* where all members of the population are included thus allowing exact measurement of variables of interest, or alternatively on a *sample* where a subset of the population is included, thereby providing only estimates of the variables of interest.

EPIDEMIOLOGICAL STUDIES

Epidemiological studies are broadly categorised into *non-observational* (or *experimental* studies) and *observational* studies. The first group includes clinical trials or intervention studies and the basic principle is that the design of the study involves deliberately changing population parameters and assessing the effect.

With this type of study an attempt is made to simplify observation by *creating* suitable conditions for the study. The second group assumes that the study does not interfere with the population characteristics.

Here, the investigator is only allowed to *select* suitable conditions for the study. *Observational studies* can be further categorised into *prospective cohort* studies and *retrospective* studies as well as *cross-sectional* studies. *Retrospective studies* include case-control and retrospective cohort studies. Another type of observational study is the *longitudinal* study which is a mix between prospective cohort and repeated cross-sectional studies.

Within the group of observational studies mixtures of study designs are common, such as for example the case-cohort study. The *case series* is a separate group of studies and frequently used in a clinical context.

This type of study typically involves dividing a group of animals into a subgroup which is being *treated* and another subgroup which is being left *untreated* and acts as a control. The decision to treat an animal or leave it untreated is typically based on random allocation - *randomisation*. After a period of time the status with respect to a response variable (e.g. disease status) is assessed for each animal.

Summary measures of the response are then compared between both subgroups. Differences in the summary values suggest the presence of an effect of the *treatment* on the response variable. Non-observational or experimental studies can be conducted as *laboratory experiments* or as *field studies* such as *clinical trials*. The latter are usually used to evaluate therapeutic or preventive effects of particular interventions, but are also useful to investigate etiologic relationships.

The *non-observational study* provides the researcher with effective control over the study situation. If the sample size is large enough a well-designed experiment will limit the effect of unwanted factors even if they are not measurable.

Control of factors other than the treatment which are likely to have an effect on disease can be achieved by using them to define homogeneous subgroups with respect to the status of these variables - *blocking* or *matching*- within which treatment is then allocated randomly.

The possibility to have excessive control over the study situation can become a weakness of the non-observational approach as it may not be representative of the real situation in the biological system anymore.

Clinical trials are considered the method of choice for investigation of causal hypotheses about the effectiveness of preventive measures, and compared with the other types of field studies they can provide the strongest evidence about causality.

There is less opportunity for systematic error compared with the observational studies. Amongst their disadvantages are the following characteristics. They require large groups, are costly, bias may be introduced through selection error and the required duration can be long if disease incidence is low.

SAMPLING OF ANIMAL POPULATIONS

The main objectives of sampling are to provide data which will allows making inferences in relation about a larger population on the basis of examining a sample in relation to for example presence/absence of animal disease or other parameters of interest.

Inferences might relate to proving that disease is not present, to detecting presence of disease or establishing the level of disease occurrence.

The objective could also be to describe levels of milk production in a population of dairy cattle or more generally provide a descriptive analysis of an animal production system for example in an African country.

DATA SOURCES

Any *data sources* for epidemiological analyses have to be evaluated with respect to their *completeness, validity* and *representativeness*. The data can be collected as part of *routine data collection* which includes laboratory submissions, disease surveillance programmes, industry- or farm/ bureau based data recording systems and abattoirs.

More recently, *structured data collection* has been found to provide a more effective way for regular monitoring of disease / production. And finally, data can be collected as part of *epidemiological studies*. Data which is based on *laboratory submissions* is useful for detecting disease.

It can become the basis of case series and case control studies. It does not provide sufficient data to allow prevalence estimation, because the enumerator and denominator are likely to both be biased.

In isolation, laboratory submissions do not provide information about causation !!!! They are also not useful for evaluation of therapies or economic effects.

The data collection process can include the whole population of interest (=*census*) or it can be restricted to a *sample*. The latter has the advantage over the *census* that results can be obtained more quickly.

A sample is less expensive to collect, and sample results may be more accurate as it is possible to make more efficient use of resources. In addition, *probability samples* result in probability estimates which allow inferences to be used for other populations.

Heterogeneity in the results can be reduced by targeted sampling of particular sub-groups within the population. Involvement of the whole population such as is necessary for a census may not be possible due to logistic or administrative problems, so that sampling becomes the method of choice. The sampling process can be described using the following terminology.

The *target population* represents the population at risk. The population effectively sampled is called the *study population*. Frequently, the target population is not completely accessible, so that it differs to a possibly unknown extent from the study population.

It is then necessary to use common sense judgement in order to assess the representativeness of the study population in relation to the target population.

The *sampling frame* lists all sampling units in the study population, and is an essential requirement for probability sampling. *Sampling units* are the individual members of the sampling frame.

INTERPRETATION OF DIAGNOSTIC TESTS

UNCERTAINTY AND THE DIAGNOSTIC PROCESS

The duties of the veterinary profession include "to maintain and enhance the health, productivity and well-being of all animals" and "to prevent and relieve animal suffering". In order to fulfill this duty, the veterinarian has to be able to diagnose disease or production problems as well as identify possible causes.

Diagnosis is the basis for a decision, such as whether to treat (or implement a programme) or to do nothing, to further evaluate, euthanase or to wait. The tools which the veterinarian uses to come to a diagnosis include factùal knowledge, experience, intuition as well as diagnostic tests. Correct use of these four mechanisms maximises the probability of a correct diagnosis. The uncertainty with regard to the effect of a treatment on a patient's health made the ancient Greeks call medicine a *stochastic art*.

Clearly, the main task of any veterinarian is to deal with the uncertainty of both, diagnosis and the outcome of treatment. It has been shown in studies of the medical profession that fear of personal inadequacy and failure in reacting to this uncertainty is a common characteristic among physicians.

This has become an even more important problem as our society becomes increasingly specialized and technological, relying on science rather than religion or magic to explain uncertainties.

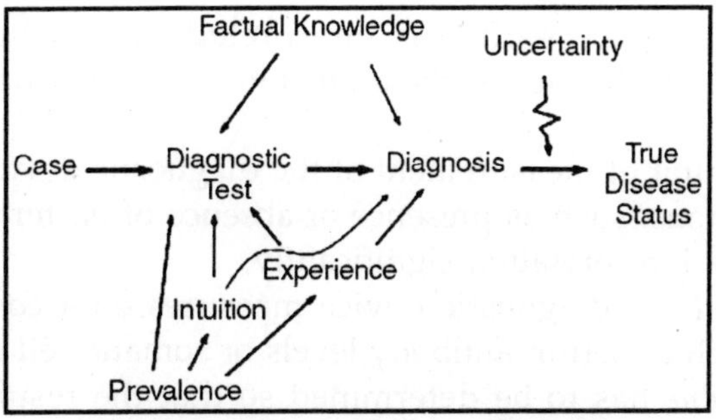

Fig. Factors Influencing Veterinary

In this context, one should be aware of two major paradigms used to explain biological processes. The *mechanistic paradigm* assumes deterministic causation, and experiments are conducted to develop rules or laws according to which nature is thought to 'work'.

Obviously, the client in a diagnostic situation does prefer this kind of interpretation, as it increases confidence in the diagnostic and the therapeutic process.

The *probabilistic paradigm* on the other hand assumes probabilistic causation.

Diagnostic and therapeutic procedures are seen as gambles and it is recognised that the decision making process incorporates subjective judgement.

The conclusion has to be though that given our incomplete understanding of biological systems and the presence of true biological variation, in veterinary diagnosis one must be content to end not in certainties, but rather statistical probabilities.

The outcome of the *diagnostic process* is a statement as to whether an animal is considered normal or not normal. This could relate to disease or infection status as well as to productive performance or quality of life from an animal welfare perspective.

DIAGNOSTIC TESTS

The diagnostic test is a more or less objective method for reducing *diagnostic uncertainty*. As the consequential decisions are typically dichotomous (treat or do not), the outcome of the diagnostic process often is interpreted as a dichotomous variable as well, such as the animal having or not having the disease.

The unit of measurement of the diagnostic device can be dichotomous, such as presence or absence of bacteria, which facilitates interpretation significantly.

But if the diagnostic device measures on a continuous scale, such as serum antibody levels or somatic cell counts, a *cut-off value* has to be determined so that the result can be condensed into a dichotomous scale.

Given a clinical measurement on a continuous scale, the problem with any cut-off point is that it is likely to result in overlap between healthy and diseased individuals with regard to test results.

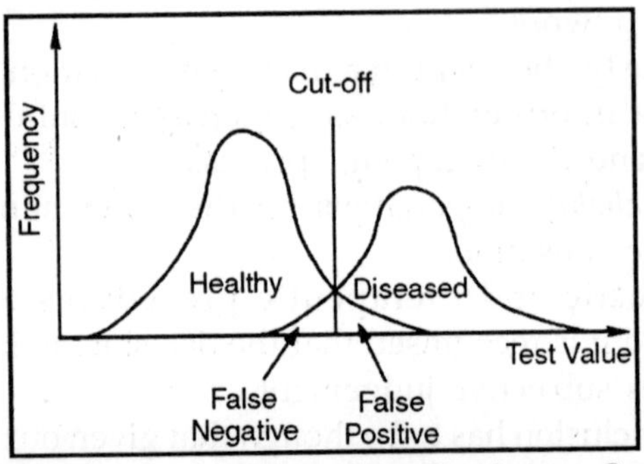

Fig. Test Result Measured on Continuous Scale

The consequences of this situation are that uncertainty in addition to any other potential sources of measurement error (such as operator error) is being introduced.

It is desirable to quantify this relationship between diagnostic test result and "true" disease status so that the clinician can take account of this uncertainty when interpreting test results.

The performance of a diagnostic method can be described using the *accuracy* which refers to the closeness between test result and "true" clinical state, the *bias* which is a measure of the systematic deviation from "true" clinical state, and the *precision* or *repeatability* representing the degree of fluctuation of a test series around a central measurement.

Any evaluation of diagnostic tests needs a measure of the "true" condition of individuals to compare with which is usually called the *gold standard*.

Most of the time it is impossible to define with 100% accuracy what the true diagnosis should be.

There may also be disagreement amongst experts such as for example in the case of mastitis where the presence of the particular pathogen *or* the presence of an inflammatory response in the udder could be defined as the gold standard.

Index

V

W

Y